Zika in Focus

Maria de Fátima Viana Vasco Aragão
Editor

Zika in Focus

Postnatal Clinical, Laboratorial
and Radiological Aspects

 Springer

Editor
Maria de Fátima Viana Vasco Aragão
Centro Diagnóstico Multimagem
Mauricio de Nassau University
Recife
Pernambuco
Brazil

ISBN 978-3-319-53642-2   ISBN 978-3-319-53643-9   (eBook)
DOI 10.1007/978-3-319-53643-9

Library of Congress Control Number: 2017932927

Printed on acid-free paper

This Springer imprint is published by Springer Nature
The registered company is Springer International Publishing AG
The registered company address is: Gewerbestrasse 11, 6330 Cham, Switzerland

Contents

Zika Virus: An Overview

Maria de Fátima Viana Vasco Aragão, Arthur Cesário de Holanda, Luziany Carvalho Araújo, and Marcelo Moraes Valença

The Zika virus is an arboviral disease (i.e., transmitted by a bite from an infected arthropod such as a mosquito) from the *Flavivirus* genus of the *Flaviviridae* family, its main vector being the *Aedes aegypti* [1]. There are also reports of sexual transmission, as well as viral detection in urine [2] and tears [3].

The Zika virus was first discovered in 1947 in monkeys, at Zika forest (Uganda, Africa) [4], and human infection was identified in 1952 [5]. Throughout the years, sporadic cases were detected in different countries, but the first epidemic of Zika virus only happened in 2007, in Micronesia and the Yap

M.F.V.V. Aragão, M.D., Ph.D. (✉)
Mauricio de Nassau University, Recife, PE, Brazil

Centro Diagnóstico Multimagem, Recife, PE, Brazil
e-mail: fatima.vascoaragao@gmail.com

A.C. Holanda
Universidade Federal de Pernambuco, Recife, PE, Brazil
e-mail: arthur.c.holanda@gamil.com

L.C. Araújo, M.D.
Hospital Barão de Lucena, Recife, PE, Brazil
e-mail: luziany_cz@hotmail.com

M.M. Valença
Universidade Federal de Pernambuco, Recife, PE, Brazil
e-mail: mmvalenca@yahoo.com.br

M.F.V.V. Aragão (ed.), *Zika in Focus*,
DOI 10.1007/978-3-319-53643-9_1,
© Springer International Publishing AG 2017

Islands [6]. The second epidemic was notified in 2013, at French Polynesia [7], and the third began in Brazil, where it was initially detected in Bahia, Northeast Brazil, in March 2015 [8]. According to the World Health Organization (WHO), January 18, 2017 [9], 76 countries and territories, especially in Latin America, had reported evidence of mosquito-borne Zika virus transmission. Thirteen countries have reported evidence of person-to-person transmission on Zika virus [9].

Diagnosis of Zika virus infection is difficult because it is asymptomatic in three of four persons infected [7], and the clinical manifestations are unspecific (e.g., rash, fever, asthenia, conjunctivitis) [8]. There are two major lineages of Zika virus, the African, reported recently in Guinea-Bissau, and the Asian, reported from Asia and the West Pacific region to the Americas and Cabo Verde [10]. Neurologic complications have been related only to Asian strains after 2007 [10].

In the second half of 2015, in Pernambuco, Northeast Brazil, a significant increase in the number of congenital microcephaly cases was identified and notified to the State Health Secretary and the Brazilian Ministry of Health. The imaging studies of these cases revealed findings suggestive of congenital infection but frequently not typical of the STORCH (syphilis, toxoplasmosis, rubella, cytomegalovirus, and herpes) infections [11]. The peak of cases was in November 2015, gradually decreasing throughout the year 2016 [12] but spreading throughout Brazil, with the Southeast becoming the second region in number of cases [10]. In Pernambuco, the highest incidence was in epidemiological week 46 of November 2015 (197 suspeted cases) [13]. From that period, it evolved with reduction, with small fluctuations.

At first, many hypotheses where proposed regarding the causal factor for this congenital microcephaly epidemic, but an increasing body of evidence started to be gathered for a relationship with the Zika virus epidemic of the first semester. This hypothesis was first raised in Pernambuco [14]. In November 2015, the Zika virus was isolated from the brain and the cerebrospinal fluid of children with congenital microcephaly and identified in the amniotic fluid and placenta of mothers who had symptoms of the infection during their pregnancies [15]. Up to the arrival of the virus in Brazil, the association between Zika virus and microcephaly had not been reported. Some cases were posteriorly and retrospectively identified in French Polynesia, in November 2015 [16]. On February 1, 2016, WHO declared Zika virus a world public health emergency [17].

Currently, the relationship between the Zika virus and microcephaly is well established [18], as well as the most characteristic findings of the congenital Zika syndrome (e.g., microcephaly, arthrogryposis, ophthalmologic, and hearing abnormalities) [19–23]. However, the disease severity spectrum, that has also less severe cases (without microcephaly), has only started to be recognized.

In Brazil, up to December 31, 2016, there were 2366 cases, with the Northeast being the most affected region (1804 cases), with the highest number of cases in Pernambuco (408 cases) [24]. In Pernambuco, in epidemiological weeks n° 1 and 2 of 2017, 11 suspected cases were reported, but not confirmed yet. Compared with the same period of the previous year, there was a reduction of 91.2 % in these notifications [13].

Up to January 18, 2017, 29 countries had reported microcephaly and other central nervous system malformations [9]. After Brazil, Colombia and the United States were the countries with most cases (78 and 41, respectively) [9].

The diagnosis of congenital Zika syndrome is currently based on clinical and radiological findings, associated with laboratory exclusion of other congenital infections or hereditary conditions (i.e., TORCHs and pseudo-TORCHs) [25]. Although microcephaly is not the only manifestation, it is the most prominent of the syndrome. It is through this feature that suspected cases are investigated, based on the Brazilian government protocol [25]. In the beginning of the epidemic, the inclusion criteria for microcephaly was a head circumference ≤33 cm [25]. In December 2015, the criterion changed to ≤32 cm and, in February 2016, it changed again, now to head circumference ≤31.5 cm for girls and ≤31.9 cm for boys [25]. Currently, the criteria for microcephaly is based on the Intergrowth-21st [26].

The major imaging abnormalities found in moderate to severe cases of microcephalic children with congenital Zika virus infection are calcifications in the junction between the cortical and subcortical white matter and malformations of cortical development (pachygyria or polymicrogyria, predominantly in the frontal lobes, and simplified giral pattern) [19]. Other frequent imaging findings are ventriculomegaly; decrease in the brain, brainstem, and cerebellum volumes; enlargement of the cisterna magna and of the extra-axial subarachnoid space; corpus callosum abnormalities (hypogenesis and hypoplasia); and delayed myelination [19]. At the spinal cord, the main abnormalities are reduced thickness and reduction of the anterior nerve roots of the medullary cone [22].

Immunologic tests specific for the Zika virus are yet in the developmental and research stages, therefore, not completely trustworthy, and they are not available for use in clinical routine. Therefore, the role of pediatricians and radiologists is extremely important in recognizing the major findings of the clinical and radiological pattern identified from confirmed and probable cases of the congenital Zika syndrome. The radiologist becomes even more essential for recognition and suggestion of the diagnosis in mild cases, without microcephaly and without epidemiological history of the disease (i.e., maternal rash during pregnancy), when the disease is not suspected by the pediatrician due to unspecific signs and symptoms. This way, laboratory tests for confirmation of the disease can be performed sooner, considering the short and yet unknown window of the specific IgM test for the Zika virus.

There is no specific treatment for microcephaly [27]. Once the disease is recognized, rehabilitation must be started immediately, especially in the less severe cases, providing the possibility of a better neuropsychomotor development for the children affected and support for their families. At the moment, prevention remains the only way to control the disease, while a vaccine or a medication is not developed and the physiopathology is not understood completely. Sanitation is the best way to control the vector, preventing sites of stagnant water. In addition, the use of long-sleeved clothes and insect repellents, for example, is also recommended [28]. Women must use condoms during the entire pregnancy in all sexual relations, due to the possibility of sexual transmission [28].

Besides congenital microcephaly, neurologic complications have been found in adults. For example, in 21 countries, according to WHO, there was an increase in incidences of the Guillain-Barre syndrome and/or laboratory confirmation of Zika virus infection among Guillain-Barre syndrome cases [10]. Myelitis and encephalitis have also been identified.

On November 18, 2016, the director general of the Emergency Committee on Zika virus, microcephaly, and other neurological disorders declared the end of the Public Health Emergency of International Concern (PHEIC) [29]. Although a decline in cases of Zika virus infection has been reported in some countries, or in some parts of countries, vigilance needs to remain high [9].

References

1. Dick GWA, Kitchen SF, Haddow AJ. Zika virus. I. Isolations and serological specificity. Trans R Soc Trop Med Hyg. 1952;46:509–20.
2. Gourinat A-C, O'Connor O, Calvez E, Goarant C, Dupont-Rouzeyrol M. Detection of Zika virus in urine. Emerg Infect Dis. 2015;21:84–6.
3. Miner JJ, Sene A, Richner JM, et al. Zika virus infection in mice causes panuveitis with shedding of virus in tears. Cell Rep. 2016;16:3208–18.
4. Dick GWA. Zika virus. II. Pathogenicity and physical properties. Trans R Soc Trop Med Hyg. 1952;46:521–34.
5. Macnamara FN. Zika virus: a report on three cases of human infection during an epidemic of jaundice in Nigeria. Trans R Soc Trop Med Hyg. 1954;48:139–45.

6. Duffy MR, Chen T-H, Hancock WT, et al. Zika virus outbreak on Yap Island, Federated States of Micronesia. N Engl J Med. 2009;360:2536–43.
7. Cauchemez S, Besnard M, Bompard P, et al. Association between Zika virus and microcephaly in French Polynesia, 2013–15: a retrospective study. Lancet Lond Engl. 2016;387:2125–32.
8. Campos GS, Bandeira AC, Sardi SI. Zika Virus Outbreak, Bahia, Brazil. Emerg Infect Dis. 2015;21:1885–6.
9. World Health Organization Situation Report. Zika virus, microcephaly, Guillain-Barre syndrome. 20 Jan 2017. http://apps.who.int/iris/bitstream/10665/253604/1/zikasitrep20Jan17-eng.pdf. Accessed 28 Jan 2017.
10. World Health Organization Situation Report. Zika virus, microcephaly, Guillain-Barre syndrome. 20 Oct 2016. http://apps.who.int/iris/bitstream/10665/250590/1/zikasitrep20Oct16-eng.pdf?ua=1. Accessed 26 Oct 2016.
11. Barkovich AJ, Raybaud C. Pediatric neuroimaging. Philadelphia: Lippincott Williams & Wilkins; 2012.
12. França GVA, Schuler-Faccini L, Oliveira WK, et al. Congenital Zika virus syndrome in Brazil: a case series of the first 1501 livebirths with complete investigation. Lancet Lond Engl. 2016;388:891–7.
13. Secretaria Executiva de Vigilância em Saúde (Pernambuco, Brazil). Informe Técnico – n° 01/2017. Síndrome congênita associada à infecção pelo vírus Zika (SCZ). 2017. https://media.wix.com/ugd/3293a8_e2103a624ad04565ab331897daff4dfd.pdf. Accessed 23 Jan 2017.
14. Brito C. Zika virus: a new chapter in the history of medicine. Acta Med Port. 2015;28:679–80.
15. Oliveira Melo AS, Malinger G, Ximenes R, Szejnfeld PO, Alves Sampaio S, Bispo de Filippis AM. Zika virus intrauterine infection causes fetal brain abnormality and microcephaly: tip of the iceberg? Ultrasound Obstet Gynecol. 2016;47:6–7.
16. Besnard M, Lastere S, Teissier A, Cao-Lormeau V, Musso D. Evidence of perinatal transmission of Zika virus, French Polynesia, December 2013 and February 2014. Euro Surveill Bull Eur Sur Mal Transm Eur Commun Dis Bull. 2014;19.
17. Tavernise S, McNeil DJ. Zika virus a global health emergence, WHO says. 2016.
18. Rasmussen SA, Jamieson DJ, Honein MA, Petersen LR. Zika virus and birth defects—reviewing the evidence for causality. N Engl J Med. 2016;374:1981–7.

19. de Fátima Vasco Aragão M, van der Linden V, Brainer-Lima AM, Coeli RR, Rocha MA, Silva PS dAM, Carvalho MDCG dB, Linden A v d, Holanda AC d, Valenca MM. Clinical features and neuroimaging (CT and MRI) findings in presumed Zika virus related congenital infection and microcephaly: retrospective case series study. BMJ. 2016;353:i1901.

20. M de C L, Muniz LF, S da S CN, van der Linden V, RCF R. Sensorineural hearing loss in a case of congenital Zika virus. Braz J Otorhinolaryngol. 2016. doi:10.1016/j.bjorl.2016.06.001.

21. Ventura CV, Maia M, Ventura BV, et al. Ophthalmological findings in infants with microcephaly and presumable intra-uterus Zika virus infection. Arq Bras Oftalmol. 2016;79:1–3.

22. Linden V v d, Filho ELR, Lins OG, et al. Congenital Zika syndrome with arthrogryposis: retrospective case series study. BMJ. 2016;354:i3899.

23. Schuler-Faccini L, Ribeiro EM, Feitosa IML, et al. Possible association between Zika virus infection and microcephaly—Brazil, 2015. MMWR Morb Mortal Wkly Rep. 2016;65:59–62.

24. Ministério da Saúde (Brazil). Informe epidemiológico n° 57. Monitoramento dos casos de microcefalia no Brasil. 2016. http:// portalsaude.saude.gov.br/images/pdf/2017/janeiro/12/Informe-Epidemiologico-n57-SE-52_2016-09jan2017.pdf. Accessed 28 Jan 2017.

25. Ministério da Saúde (Brazil). Protocolo de vigilância e resposta à ocorrência de microcefalia e/ou alterações do sistema nervoso central (SNC). 2016. http://portalsaude.saude.gov.br/images/ pdf/2016/janeiro/22/microcefalia-protocolo-devigilancia-e-resposta-v1-3-22jan2016.pdf. Accessed 16 May 2016.

26. Ministério da Saúde (Brazil). Protocolo de vigilância e resposta à ocorrência de microcefalia e/ou alterações do sistema nervoso central (SNC). 2016. http://combateaedes.saude.gov.br/images/ sala-de-situacao/Microcefalia-Protocolo-de-vigilancia-e-resposta-10mar2016-18h.pdf. Accessed 19 Dec 2016.

27. Falcao MB, Cimerman S, Luz KG, et al. Management of infection by the Zika virus. Ann Clin Microbiol Antimicrob. 2016;15:57.

28. Center for Disease Control and Prevention. Key Messages - Zika Virus Disease. 18 Jan 2017. https://www.cdc.gov/zika/pdfs/ zika-key-messages.pdf. Accessed 30 Jan 2017

29. World Health Organization Situation Report. Zika virus, microcephaly, Guillain-Barre syndrome. 24 Nov 2016. http://apps.who. int/iris/bitstream/10665/251648/1/zikasitrep24Nov16-eng. pdf?ua=1. Accessed 3 Jan 2017.

Zika Virus: History and Infectology

Carlos Alexandre Antunes de Brito

Introduction

Zika is an arbovirus of the *Flaviviridae* family that includes other human pathogens such as the dengue virus, the yellow fever virus, and the West Nile virus. The Zika virus was first identified in the Zika Forest in Uganda in 1947 with some sporadic cases of infection in humans.

Phylogenetic studies point to two African lineages and one Asian lineage that in 2007 caused the first epidemic outbreak in Micronesia, on the island of Yap. The virus spread, and later in 2013 in French Polynesia, a major epidemic outbreak occurred, being estimated that 11% of the population had been infected. In 2014, the Asian strain of the virus arrived in the Americas with some cases on the Chilean territorial Easter Island, continuing on to the northeast of Brazil in early 2015. At that time, there was an outbreak of an exanthematous disease characterized by precocious and prickly exanthem, slight fever, arthralgia, articular edema, and conjunctivitis. The autochthonous transmission of the Zika virus

C.A.A. Brito, M.D., Ph.D.
UFPE-Federal University of Pernambuco, Recife, PE, Brazil
e-mail: cbritoc@gmail.com

M.F.V.V. Aragão (ed.), *Zika in Focus*,
DOI 10.1007/978-3-319-53643-9_2,
© Springer International Publishing AG 2017

in Bahia and Rio Grande do Norte was confirmed through (RT-PCR) viral isolation.

Zika virus is transmitted by mosquitoes of the *Aedes* genus, including *Aedes aegypti* and *Aedes albopictus*, which are also responsible for the transmission of dengue and chikungunya. Studies have shown other forms of transmission, by blood transfusion, sexual contact, and perinatal transmission.

Historical Evolution

Although the identification of the Zika virus occurred in 1947, the infection continued to be the subject of few published studies due to the frequency of asymptomatic and oligo-symptomatic presentations before 2007, with cases occurring sporadically and without reports of severe forms [1–3].

Some data were initially published based on prevalence studies and a modicum of samples. In 1984, in Uganda, 6% of 132 adult samples had detectable Zika antibodies [4]. In Nigeria, in 1979, 31% of 189 serum bank samples were positive by inhibition of hemagglutination for Zika and in 38% were positive by the detection of neutralizing antibodies [5].

The first reported significant epidemic outbreak occurred in 2007 on Yap Island, which is part of the Federated States of Micronesia in the Pacific Ocean, where physicians observed an outbreak of a disease characterized by skin rash, conjunctivitis, subjective fever, arthralgia, and arthritis. The outbreak was confirmed 3 months later with the detection of Zika virus by RT-PCR in 10 of 71 samples. An attack rate was estimated as 14.6 per 1000 inhabitants; there is a seroprevalence of 72% for the population over 3 years of age, and only 18% were symptomatic among those infected [6].

However, it was from a major outbreak in French Polynesia, in 2013, that we have the most robust data on the disease with a change of pattern, characterized by high attack rates and reports of the onset of neurological complications in adults, which up to then had gone unnoticed.

A serological study prior to the epidemic of 2013 among blood bank donors showed the previously nonexistent circulation of this virus in the country [7]. During the epidemic, 8262 suspected cases were reported in the sentinel units. Of a total of 746 samples tested, 396 (53.1%) were confirmed by RT-PCR. Over 29,000 cases were estimated with an attack rate of 10% [8, 9]. In this epidemic, 39 cases of Guillain-Barre syndrome (GBS) emerged after the epidemic, potentially associated with Zika virus, but without viral isolation.

In Brazil, the first reports of suspected Zika cases occurred in the northeast, at the end of 2014 with a peak in the first quarter of 2015, characterized by a large outbreak of an exanthematous disease with a clinical pattern different from that of dengue. The hypothesis that it was Zika only came to be confirmed in April 2015 in 8 of 25 samples from suspected cases in Bahia and later in 8 of 21 cases in Rio Grande do Norte [10–12].

Just like what happened in French Polynesia in the first semester of 2015, an increase in neurological cases was observed in the region [8]. Confirmation of the association occurred in the state of Pernambuco, where seven patients presented positive results for Zika by RT-PCR and viral isolation, of which six were in serum samples and one in CSF. Of these cases, four had a diagnosis of GBS, two of acute disseminated encephalomyelitis (ADEM), and one of meningoencephalitis. Subsequently, seven other countries in the Americas reported an increase in the number of GBS cases [13].

In October 2015, physicians from the northeast of Brazil reported an alarming increase in cases of microcephaly, mainly in the states of Pernambuco, Sergipe, and Rio Grande do Norte. During this period a hypothesis is put forward by some specialists of the northeast that microcephaly is associated with perinatal infection by Zika virus, based on the following aspects: (a) there is a high rate of dispersion, suggestive of transmission associated to the vector, characterized by a large number of cases, coming from various cities, occurring within a short period of time, (b) 70% of the mothers reported an exanthematous disease in the first trimester compatible with Zika, (c) the neurological accompaniment of the adult detected in that year would suggest a neurotropism of the virus, (d) the neuroimaging tests would suggest congenital infection, (e) the other congenital infections associated with microcephaly were discarded, and (f) there is a temporal relationship between the Zika outbreak at the beginning of 2015 and the time of the appearance of microcephaly [12, 14, 15].

The detection of a large number of microcephaly cases compared to previous historical records in the state of Pernambuco made Brazil's Ministry of Health on October 27, 2015, to communicate the fact to international authorities. In November, the Ministry of Health declared a national public health emergency and sent to all state health secretariats guidelines on the process of notification, surveillance, and assistance to pregnant women and newborns stricken by microcephaly [16].

The RT-PCR examinations and viral isolation in the CSF and blood samples tested for the first cases were negative. The justification for the negative results could be related to the fact that Zika infection occurring in the first trimester of pregnancy would not allow identification of the virus in live births months after exposure, although this isolation in other infections such as cytomegalovirus may be possible for months after birth.

The first confirmation came on November 17, when a specialist in fetal medicine in Paraíba identified by RT-PCR the Zika virus infection in amniotic fluid in two pregnant women in their fifth month of pregnancy, whose fetuses presented microcephaly. The viral genome was identified as being from the Asian lineage [17, 18].

On November 28, 2015, the Brazilian Referral Center in Research, Instituto Evandro Chagas, detected the presence of the virus in blood and tissue exams of two stillborn infants with microcephaly, reinforcing and confirming the association of congenital malformation with the Zika virus [16].

After the Brazilian alert, health authorities in French Polynesia have been investigating retrospectively cases of brain malformations in fetuses and newborns after the Zika epidemic in 2014 and have identified at least 17 recorded cases [19]. In March 2016, the amniotic fluid stored from four of these cases confirmed the presence of the virus by RT-PCR [20].

Other works were published in subsequent months definitively confirming the association. In December 2015, tissue samples from two stillbirths and two abortions with microcephaly, from Rio Grande do Norte, in the northeast of Brazil, were sent to the Centers for Disease Control and Prevention (CDC) for histopathological evaluation and laboratory tests for suspicion of Zika virus infection. All four mothers presented clinical signs of Zika virus infection, including fever and exanthem during the first trimester of pregnancy. Samples from the brain and other autopsy tissues of the two newborns, the placenta of one of the newborns, and conception products from the two spontaneous abortions were positive for Zika by RT-PCR and immunohistochemistry [21].

A case of a pregnant woman contaminated in Brazil and whose fetus was diagnosed with microcephaly after returning to her country of origin occurred in Slovenia. There was interruption of pregnancy, and the presence of the virus was identified in the brain tissue by RT-PCR [22].

The analysis of the CSF from the first 42 cases of microcephaly in Pernambuco, with diagnosis of probable congenital infection associated with Zika, detected the presence of specific IgM for Zika virus. As IgM does not penetrate the placental barrier, it indicates unequivocal infection of the newborn. This would be proof for reinforcing the claim that the majority of cases of the new outbreak in microcephaly that has stricken the country was due to Zika [23, 24].

In February 2016 the World Health Organization (WHO) declared the epidemic by Zika an international public health emergency and in April finally recognized the causal relationship between Zika and microcephaly. In the November

2016 WHO bulletin, there were 2311 cases of congenital syndrome associated with Zika virus infection distributed over 24 countries. 2143 of these cases were reported in Brazil. The majority of the cases in Brazil are cases with clinical diagnosis and imaging, with 417 confirmed in the laboratory as Zika virus infection [25, 26]. Most of the Brazilian cases were recorded in the northeastern region, with the state of Pernambuco having the highest number of cases with clinical and radiological confirmation (393), and a further 349 under investigation, followed by Bahia with 339 confirmed cases, Paraíba (186), Ceará (150), Rio de Janeiro (149), and Rio Grande do Norte (141).

Sexual Transmission

One of the first reports of sexual transmission occurred in 2008 when an American researcher returning from Senegal developed Zika's symptoms and his wife also presented a clinical picture of Zika, confirmed by serological tests [27]. In February 2016, the CDC reported 14 cases of suspected sexual transmission in the United States, three of them with laboratory confirmation [28]. Other reports in various countries were confirmed [29–31].

Recent publications reinforce a significant virus excretion in sperm [32–34]. In France, researchers detected a viral load 100,000 times greater in semen than in blood and urine, 2 weeks after the onset of symptoms [33]. The detection of virus by RT-PCR has already been possible in a sample of semen 93 days after acute infection. Because of new evidence, the CDC and the UK government have recommended that partners from countries with Zika virus transmission should refrain from sexual intercourse or use condoms for at least 8 weeks after returning on the part of those who did not develop symptoms and 6 months for those with clinical confirmation of infection. For pregnant women with partners from these regions, the recommendation is of safe sex throughout pregnancy [35, 36]. Recently the virus was also transmitted from a symptomatically infected woman to a male sexual partner, and the RNA of the virus was detected in vaginal fluids 3 days after the onset of symptoms and in the cervical mucus up to 11 days after the onset of symptoms, amplifying the recommendation of the CDC that includes pregnant women with female sexual partners [37–39].

Breastfeeding

The virus was detected in the breast milk of two perinatal women who had Zika infection, but viral replication was not detected making transmission via this route unlikely. The WHO reinforces that the benefits of breastfeeding outweigh any potential risk of Zika virus transmission through breast milk in the light of available evidence [40].

Transfusion

Transmission by transfusion has been demonstrated in a study conducted in French Polynesia among 1505 blood donor samples with the detection of the presence of Zika virus by RT-PCR in 42 (3%) samples, of which 11 (26%) reported having had Zika 3–10 days after donation [41].

Clinical Spectrum

Classic Form

In symptomatic patients, the most frequent manifestations are maculopapular rash, slight fever, arthralgia, myalgia, headache, and non-purulent conjunctive hyperemia. In the Yap Island outbreak in 2007, the analysis of 31 cases showed rash to be the predominant symptom present in 90% of the patients (average duration of 6 days with a variation of 2–14 days), and fever was less frequent, affecting 65% of patients, including not only measured fever but also stated fever. Arthritis and arthralgia were frequent, affecting 65% of patients (average duration of 3.5 days ranging from 1 to 14 days). Non-purulent conjunctivitis was present in 55% of cases, myalgia and headache in around 45%, and recurring pain in 39% [6, 8].

Fever is typically of low intensity (<38.5 ° C) or absent, often only subjectively reported by the patient, occurring intermittently throughout the first 48 h of the disease, but not persistent as referred to in the classic forms of dengue. Elevated fever may occur less frequently and in neurological cases.

Skin rash arises early within the first 48 h of the onset of symptoms, unlike other arboviruses such as dengue that appear after 3–4 days of fever. The rash can present different patterns and intensities, but is usually of the morbilliform type, starting in the face and rapidly progressing to the trunk, upper limbs, and finally lower limbs, in an additive form (Figs. 1 and 2).

In pregnant women the main manifestation is the presence of exanthema. In the study of Brazil P, involving 72 pregnant women with positive RT-PCR for Zika, exanthema was present in 100% of cases, pruritus in 96%, arthralgia/arthritis in 64%, conjunctivitis in 58%, headache in 53%, retro-orbital pain in 49%, myalgia in 41%, and edema in 35% [41].

The differential diagnosis should be made mainly with dengue, chikungunya, measles, and rubella given priority [42] (Table 1).

Neurological Manifestations

The emergence of neurological cases potentially associated with Zika virus was initially described in French Polynesia in 2013, with reports of 39 cases of GBS arising after the epidemic but without virus isolation [8, 9].

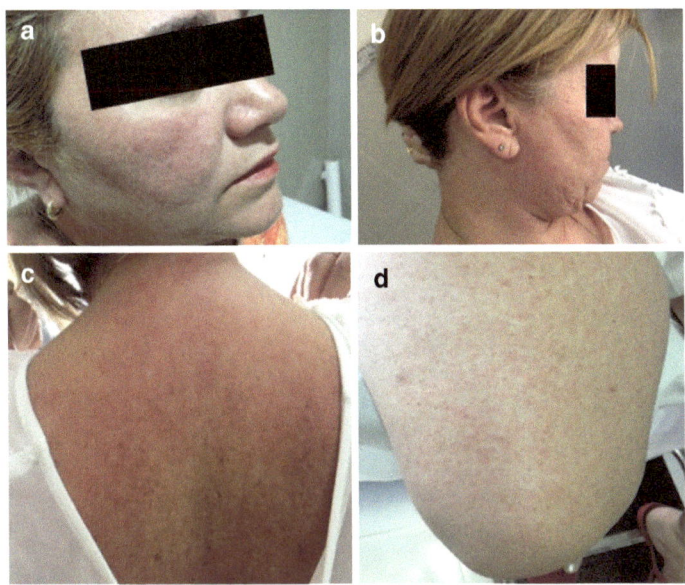

FIGURE I Clinical manifestations in Zika virus. (Photos **a** and **b**): Patients with face rash. There is discrete edema in the face referred by the patients. (Photo **c**) Rash on back. (Photo **d**) Rash in lower limb

In northeastern Brazil, an increase in neurological cases was noticed from April 2015 [12]. Confirmation of the association occurred in the state of Pernambuco, where seven patients—six in serum samples and one in CSF—presented positive results for Zika by RT-PCR and viral isolation. Of these cases, four had a diagnosis of GBS, two of acute disseminated encephalomyelitis (ADEM), and one of meningoencephalitis. Another 70 neurological cases are under investigation with a view to detailing the outbreak [12, 37]. Seven other countries in the Americas reported an increase in the number of GBS cases [14].

Recently, 41 neurological cases described in 2013 with GBS from French Polynesia had serological confirmation in stocked serum samples. The pattern of acute motor axonal neuropathy predominated among the cases (74%). The mean

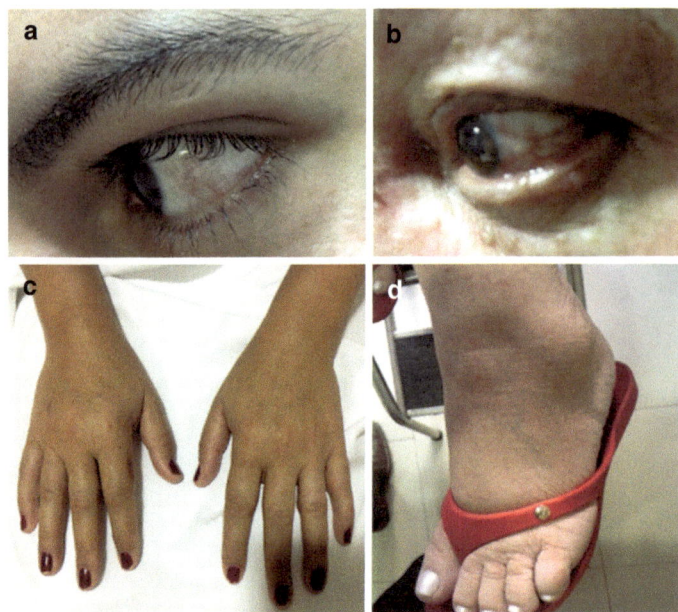

FIGURE 2 Clinical manifestations in Zika virus. (Photos **a** and **b**) Patients with non-purulent conjunctivitis (mild ocular hyperemia). (Photo **c**) Wrist and finger edema. (Photo **d**) Ankle edema

time between infection and development of neurological symptoms was 6 days [43]. A recent publication questioned the interpretation of the serological tests performed and claimed that the results do not allow concluding within a laboratory that it is Zika, due to potential cross-reaction with dengue [44].

The first report of the pathogenesis of the virus was described by Dick GW et al. in an animal model published in 1952. The inoculation of the Zika virus in mice triggered motor weakness and limb paralysis with viral replication detected in the brain tissue, suggesting a neurotropic virus [45].

The short time between the virus and the onset of disease reported in French Polynesia [43] for GBS cases may suggest

TABLE I Differential features of disease caused by Zika virus and other arboviruses

Signs and symptoms	Dengue	Zika	Chikungunya
Fever Duration	>38 °C 4–7 days	Without fever or subfebrile (≤38 °C) 1–2 days subfebrile	High fever >38 °C 2–3 days
Skin spots	Appears on the fourth day	Appears on the first or second day	Appears 2–5 days
Frequency	30–50% of the cases	90–100% of the cases	50% of the cases
Muscle pain (frequency)	+++	++	+
Joint pain (frequency)	+	++	+++
Intensity of joint pain	Mild	Mild/moderate	Moderate/intense
Swelling of the joint	Rare	Frequent and mild intensity	Frequent and moderate to intense
Pink eye	Rare	50–90% of the cases	30%
Headache	+++	++	++
Hypertrophy of the lymph node	+	+++	++
Hemorrhagic dyscrasia	++	Absent	+
Death risk	+++	+[a]	++
Neurological involvement	+	+++	++

(continued)

TABLE I (continued)

Signs and symptoms	Dengue	Zika	Chikungunya
Leukopenia	+++	+++	+++
Lymphopenia	Uncommon	Uncommon	Often
Thrombocytopenia	+++	Absent	++

[a]There may be death risk in cases such as neurological GBS resulting in Zika or for children with severe congenital malformations [42]

a direct neuropathic effect of the virus, in addition to immune-mediated effect with synthesis of antibodies that damage the peripheral nerve and spinal roots, and that it occurs weeks after the infection. These and other questions are to be clarified in future studies of immunopathogenesis.

Microcephaly

Microcephaly is an important neurological sign related to reduced brain volume associated with alterations in neuro-psychomotor development. It is a rare event, present in 0.56% of the child population. It may be present at birth being classified as primary, related to genetic or secondary causes, related to etiologic agents that cause injury to the developing fetal brain, such as infections of the TORCH group and teratogenic agents. Microcephaly associated with Zika would be classified as congenital, secondary to infection by the virus during the first months of gestation.

Although there is no uniformity in the literature regarding the definition of microcephaly, it is generally defined as the cephalic perimeter smaller than 2DP from the mean for gestational age and sex and considered severe when smaller than 3DP. The cephalic perimeter (CP) should be measured with a non-extensible measuring tape, firstly at the height of the supraorbital arches and then at the larger prominence of the occipital bone, within the first 24 h of life.

In October of 2015, due to the unexpected increase in cases of microcephaly in newborns, the Ministry of Health opted at the outset to adopt a 33-cm average for the cephalic perimeter to include a greater number of newborns in the research. This course of action was established in the face of the nonuse, in most Brazilian nurseries, of the CP curve for gestational age and sex. Subsequently, as many newborns were unnecessarily screened, the WHO criteria began to be adopted, i.e., boys and girls with less than 32 cm in head circumference. This corresponds to a percentage of 2.6 for boys and 5.6 for girls, both on the CP chart adopted by the WHO as well as by the CDC, bringing it closer to the international definition of microcephaly.

For the preterm newborn, the cephalic perimeter is less than −2 standard deviations, by the Fenton curve for gestational age and sex.

However, as of March 9, 2015, the Ministry of Health, following the recommendations of the WHO in order to standardize the references for all countries, began to define microcephaly in full-term newborns as a CP less than 31.5 cm in girls and 31.9 in boys. For premature infants, the InterGrowth chart began to be used which has the gestational age of the newborn as a reference.

It is important to emphasize that the diagnosis of microcephaly should be confirmed with imaging tests, such as transfontanellar ultrasonography or cranial tomography.

Congenital Zika

The congenital Zika syndrome is a disease that has not yet been fully clarified and has recently been described, associated with the widespread epidemic of microcephaly in Brazil. Its clinical and laboratory aspects have been defined so far, through already published case reports. However, it is not known if there is a broader spectrum of clinical aspects than those already found, and microcephaly is only the tip of the iceberg of this syndrome.

Clinical Aspects

1. *General*—Most newborns are in full term, with a rate of prematurity similar to the general population. A greater tendency among female children (around 60%) is noticeable.

2. *Birth conditions*—Although they are newborns (NBs) with significant malformation of the nervous system, the need for some resuscitation procedure in the delivery room is similar to the general population. Children who are born with associated malformations, in addition to microcephaly, need more support at birth.

3. *Birth weight*—In relation to birth weight, there is a higher percentage of newborns with birth weight less than 2500 g, considered to be underweight, reaching up to 30%. It is known that the head of the NB weighs 1/4 its weight and this should influence the birth weight; however, the possibility of intrauterine viral infection having an influence on fetal growth, similar to other congenital infections, cannot be ruled out.

4. *Microcephaly*—It is observed that up to 80% of the microcephaly of these newborns are considered severe microcephaly, with CP < than 3DP on average for the gestational age and sex. The mean CP varies between 27 and 29 cm, which suggests that the virus is responsible for a significant brain injury with expressive cortical atrophy. The anterior fontanelle is small and the posterior one is sometimes closed. In the most severe forms, there is a redundancy of the skin of the scalp forming folds, by the continuous growth of the skin with deceleration of cranial growth.

5. *Neurological examination*—Neurological abnormalities are the main clinical manifestations and include muscle tone alterations with hypertonia and spasticity. They are irritable children with hyperreflexia and tremors. Seizures are more frequent after the neonatal period. Although patients are discharged from the hospital on an oral diet, it

is observed that some patients present dysphasia due to the lack of coordination in sucking/swallowing.

6. *Visual alterations*—Microphthalmia, cataracts, and intra-ocular calcifications may be found. Studies report up to 30% of patients with suspected Zika congenital infection who present alterations in indirect ophthalmoscopy such as corneal atrophy and pigment deposition in the macular area and hypoplasia of the optic disc, pallor, and papillary excavation. It is important to note that TORCH infections were discarded in these patients. The authors emphasized that even in cases of suspected congenital infection by Zika virus without microcephaly, visual investigation should be performed.

7. *Hearing changes*—There are still no studies reporting changes in hearing in neonates with congenital Zika. The examination of the evoked otoacoustic emissions (EOAE) is recommended as the first examination; however, the presence of microcephaly is an indicator of risk for loss of hearing, and in these cases the brain stem-evoked response audiometry (BERA) should be the chosen exam because of the greater prevalence of retrocochlear auditory losses that are unidentifiable in the EOAE.

8. *Associated malformations*—In 15% of the patients, other malformations are found, such as arthrogryposis of the hip and limbs, congenital twisted feet, and some rare cases of unilateral diaphragmatic paralysis. Usually these malformations occur together, accompanying the microcephaly. The results are two forms of presentation of congenital Zika, one with only microcephaly and another, more severe form, with microcephaly and these malformations associated.

Unlike other congenital infections mainly of the TORCH group, no hepatosplenomegaly or hematological alterations were observed, strongly suggesting the neurotropism of the virus to the nervous system.

Laboratory Diagnosis

The definitive diagnosis of congenital infection by Zika virus depends on the isolation of the virus in viral culture, but this depends on the sample analyzed and the viral load.

Rapid and accurate diagnosis can be done through RT-PCR to the Zika virus that detects fragments of the viral genome. Viremia is short lived, lasting between 7 and 10 days, so samples should be collected during this time. In cases of pregnant women with exanthematous viral disease, the RT-PCR for Zika virus should be collected and if positive be defined as a confirmed case under risk of fetus with microcephaly exposure to Zika virus. The pregnant woman should initiate adequate accompaniment according to Ministry of Health protocol.

Samples taken from urine, semen, and amniotic fluid are more positive because they have a higher viral load and can be positive up to 60 days of infection. The reports of cases that confirmed intrauterine infection by Zika virus were done with RT-PCR of amniotic fluid and viral particles present in fetal tissues.

After the acute phase of the disease, when the RT-PCR for Zika virus has already negated it, the option is to use serological tests. The IgM for Zika virus (ELISA) is positive within the first 2 weeks after the acute phase and may last up to 2–6 months. The major limitation of this examination is cross-reactivity with other flaviviruses, such as dengue and chikungunya, especially in endemic areas. In these cases, neutralizing antibodies are used for diagnostic confirmation to attest recent infection by Zika virus.

In 22 infants with suspected Zika virus congenital infection associated with microcephaly, CSF samples were analyzed and 21 tested positive for IgM by Zika virus, confirming intrauterine transmission infection and virus neurotropism.

Diagnosis by Neuroimaging

The diagnosis of congenital Zika syndrome begins in the prenatal period through laboratory tests and is complemented by obstetric ultrasonography. This should be requested for all pregnant women with a history of exanthematous viral disease in endemic areas. A suspicious case is considered when the obstetric ultrasound of the patient shows a fetus with cranial circumference (CC) less than two standard deviations (<2 DP) below the mean for gestational age or with alteration in the central nervous system (CNS) suggestive of congenital infection, as long as other congenital infections are discarded. The most commonly found alterations are bilateral ventriculomegaly, intracranial calcifications, and groove alterations.

After birth, transfontanellar ultrasonography reveals the same alterations seen in fetal ultrasonography as well as cerebellar hypoplasia, cortical atrophy, pachygyria, and even more severe cases of neuronal migration disorders, lissencephaly. Of the 25 cases of transfontanellar ultrasound investigation, only in one was an adequate window not achieved.

Computerized tomography (CT) of the skull confirms these cerebral malformations. However, it visualizes better the cerebellar hypoplasia and generalized sulcation alterations. Other findings include brain stem hypoplasia and white matter hypodensity. These findings strongly suggest that the brain injury caused by the virus occurred in the initial stages of fetal brain development in the early months of gestation.

It is important to emphasize that CT exposes the NB to a large radiation load, requires sedation, and is an expensive exam in the health system. Therefore, as the first neuroimaging examination of the newborn with suspected congenital Zika, we suggest transfontanellar ultrasonography; and CT should be reserved for some cases or in the follow-up of these children according to indication.

Current Situation in Pernambuco: IMIP

The Technical Report No. 63 of the Executive Secretariat of Health Surveillance/State Department of Health (Cases of Microcephaly and Update: March 14, 2016) registers that 703 cases of microcephaly associated with Zika were reported, with 48% in the city of Recife. Of all the cases reported, 114 (16%) were from the Instituto de Medicina Integral Prof. Fernando Figueira (IMIP).

Currently 90 cases of congenital Zika syndrome, considered as the presence of microcephaly plus suggestive alterations in neuroimaging (transfontanellar US or cranial tomography), are being followed up at the IMIP. Twenty-one of these present positive IgM for Zika virus in LCR, confirming the diagnosis. Table 2 shows the clinical and laboratory characteristics of these children.

Most newborns are in full term, female, and with low birth weight and are born with good vitality. It is worth noting the large number of cases with an average CP <3 SD for sex and gestational age (86%). One-fourth of newborns present auditory and ophthalmological alterations.

The neurological examination was altered in all cases. In one NB, associated malformations, arthrogryposis, crooked foot, and unilateral diaphragmatic paralysis were observed. All were discharged from mixed breastfeeding. The only death was the NB with associated malformations on the tenth day of life due to respiratory failure.

The brain alterations found in the transfontanellar US and CT are shown in Table 3. All infants presented altered transfontanellar US and altered CT.

Treatment

To date, there is no specific treatment for congenital Zika syndrome. Treatment is based on supportive actions to aid in the growth and development of these children. Specific complications such as neurological, orthopedic, respiratory, and other complications will be managed by various specialists in rehabilitation services.

Zika and Breastfeeding

There is no contraindication for breastfeeding in newborns with congenital Zika syndrome due to the fact that the mother no longer has viremia and the disease is already established.

TABLE 2 Clinical and laboratory characteristics of the NB

Characteristics of the newborns	$N = 21$ (%)
Sex	
Female	14 (67)
Male	7 (33)
Gestational age	$X = 37.6 \pm 3.1$
Term	16 (76)
Preterm	5 (24)
Weight	$X = 2411 \pm 633$
<2500 g	12 (57)
>2500 g	9 (43)
Apgar ($\dot{X}$)	
First minute	9
Fifth minute	9
Weight for age adjustment	
SGA	12 (57)
AGA	9 (43)
Brain perimeter	
Microcephaly ≥3DP	18 (86)
Microcephaly <3DP	3 (14)
Fundoscopy (17)	
Alterated	4 (24)
Normal	13 (76)
Not done	4
Otoacoustic emissions (EOAE) (15)	
Alterated	4 (27)
Normal	11 (73)
Not done	6
BERA (16)	
Alterated	3 (19)
Normal	13 (81)
Not done	5

TABLE 3 Neuroimaging alterations

	Transfontanellar US $n = 20$ (%)	Head CT $n = 16$ (%)
Not performed	1 (windowless)	5
Ventriculomegaly	16 (80)	16 (100)
Intracranial calcifications	17 (85)	15 (94)
Hypoplasia of posterior fossa structures	1 (5)	7 (44)
Groove alterations	2 (10)	14 (87)

In relation to the mother presenting Zika disease during breastfeeding, Zika virus was isolated in maternal milk from two mothers with active Zika disease. However, the virus was not identified in viral culture. To date, there has been no reported transmission of Zika virus through breast milk, so due to the innumerable benefits of breastfeeding, especially in countries where Zika virus is endemic, the WHO maintains its recommendations and does not contraindicate breastfeeding in these cases.

Accompaniment

The children should be accompanied by means of childcare and also referred to specialized care and rehabilitation services by various healthcare professionals—neurologist, orthopedist, physiotherapist, speech therapist, occupational therapist, and psychologist.

The early stimulation programs for these children should be initiated as soon as the child is stable, since the first 2 years of life is the period of greatest brain development and with neuronal plasticity strongly present, creating opportunities to establish the greatest possible functional gain in motor, cognitive, and language areas.

The involvement of parents and family is essential, considering that the social environment is the richest in stimuli for the child. It is the responsibility of the multi-professional team to guide and promote the contribution of the family and community in order to enhance the development of these children in the future.

Acknowledgment I would like to thank Luziany Carvalho Araújo for her inestimable help.

References

1. Dick GWA, Kitchen SF, Haddow AJ. Zika VIRUS (I). Isolations and serological specificity. Trans R Soc Trop Med Hyg. 1952;46(5):509–30.
2. Simpson DIH. Zika virus infection in man. Trans R Soc Trop Med Hyg. 1964;58(4):335–8.
3. Macnamara FN. Zika virus: a report on three cases of human infection during an epidemic of jaundice in Nigeria. Trans R Soc Trop Med Hyg. 1954;48(2):139–45.
4. Rodhain F, Gonzalez JP, Mercieg E, HelyncK B, Larouze B, Hannoun C. Arbovirus infections and viral haemorrhagic fevers in Uganda: a serological survey in Karamoja district, 1994. Trans R Soc Trop Med Hyg. 1994;83:851–4.
5. Fagbami AH. Zika virus infections in Nigeria: virological and seroepidemiological investigations in Oyo State. J Hyg. 1979;83(2):213–9.
6. Duffy MR, Chen T-H, Hancock TW, Powers AM, Lanciotti RS, Pretrick M, et al. Zika virus outbreak on Yap Island, Federated States of Micronesia. N Engl J Med. 2009;360:2536–43.
7. Teissier A, Roche C, Broult J, Aubry M, Paulous S, Despre P. Diseases Seroprevalence of arboviruses among blood donors in French. Int J Infect. 2015;41:11–2.
8. ECDC. European Center for Disease Prevention and Control. RAPID RISK ASSESSMENT Zika virus infection outbreak, French Polynesia. 2014. p. 1–12.
9. Polynésie Française. Surveillance de la dengue et du zika en Polynésie française. 2014. p. 6–9.
10. Campos GS, Bandeira AC, Sardi SI. Zika virus outbreak, Bahia. Brazil Emerg Infect Dis. 2015;21(10):1885–6.
11. Zanluca C, Campos V, Melo AD, Luiza A, Mosimann P, Igor G, et al. First report of autochthonous transmission of Zika virus in Brazil. Mem Inst Oswaldo Cruz. 2015;110(June):569–72.
12. Brito C. Zika virus: a new chapter in the history of medicine. Acta Med Port. 2015;28(6):679–80.

13. PAHO. Zika—Epidemiological Update. 2016. p. 1–6.
14. Becker R. Missing link: animal models to study whether Zika causes birth defects. Nat Med [Internet]. Nature Publishing Group. 2016;22(3):225–7 .Available from: http://www.nature.com/doifinder/10.1038/nm0316-225
15. Teixeira MG, Da Conceição N, Costa M, De Oliveira WK, Nunes ML, Rodrigues LC. The epidemic of Zika virus-related microcephaly in Brazil: detection, control, etiology, and future scenarios. Am J Public Health. 2016;106(4):601–5.
16. Brazil. Protocolo de Atenção à Saúde e Resposta à Ocorrência de Microcefalia Relacionada à Infecção pelo Vírus Zika. 2015. p. 1–49.
17. Melo AS, Malinger G, Ximenes R, Szejnfeld P, Sampaio SA. Bisbo de Filippis A. Physician Alert Ultrasound Obs Gynecol. 2016;47:6–7.
18. Calvet G, Aguiar RS, Melo ASO, Sampaio SA, de Filippis I, Fabri A, et al. Detection and sequencing of Zika virus from amniotic fluid of fetuses with microcephaly in Brazil: a case study. Lancet Infect Dis. 2016;16:653–60.
19. Besnard M, Mallet H. Increase of cerebral congenital malformations among newborns and fetus in French Polynesia, 2014–2015, following a Zika virus outbreak. 2015;(November):2014–5. Jouannic J-M, Friszer S, Leparc-Goffart I, Garel C, Eyrolle-Guignot D. Zika virus infection in French Polynesia. Lancet. 2015;6736(16):1051–2.
20. Martines RB, Bhatnagar J, Keating MK, Silva-flannery L, Muehlenbachs A, Gary J, et al. Notes from the field: evidence of Zika virus infection in brain and placental tissues from two congenitally infected newborns and two fetal losses—Brazil, 2015. MMWR Morb Mortal Wkly. 2016;65(06):1–2.
21. Mlakar J, Korva M, Tul N, Popović M, Poljšak-Prijatelj M, Mraz J, et al. Zika virus associated with microcephaly. N Engl J Med [Internet]. 2016;374(10):951–8.
22. Cordeiro M, Pena L, Brito CA, Gil LH, Marques ET. Correspondence Positive IgM for Zika virus in the cerebrospinal fluid of 30 neonates with microcephaly in Brazil. Lancet [Internet]. 2016;6736(16):2–3. doi:10.1016/S0140-6736(16)30253-7.
23. Cordeiro M, Broto C, Pena L, Castanha P, et al. Results of a Zika virus (ZIKV) immunoglobulin M–specific diagnostic assay are highly correlated with detection of neutralizing anti-ZIKV antibodies in neonates with congenital disease. J Infect Dis. 2016;214:1897–904.

24. WHO/PAHO. Zika suspected and confirmed cases reported by countries and territories in the Americas Cumulative cases, 2015–2016. Updated as of 17 Nov 2016.

25. Brasil. Ministério da Saúde. Monitoramento de casos de microcefalia. Informe_Epidemiologico_Micorcefalia n_ 51_ SE44_2016. pdf. http://portalsaude.saude.gov.br/images/pdf/2016/novembro/21/Informe_Epidemiologico_n_%2051_%20SE44_2016.pdf.

26. Foy BD, Kobylinski KC, Foy JLC, Blitvich BJ. Probable non-vector-borne transmission of Zika virus, Colorado, USA Emerg Infect Dis.2011;1–7.

27. Hills SL, Russell K, Hennessey M, Williams C, Oster AM. Transmission of Zika virus through sexual contact with travelers to areas of ongoing transmission–Continental United States, 2016. Morb Mortal Wkly Rep. 2016;65(8):215–6.

28. Venturi G, Zammarchi L, Fortuna C, Remoli ME, Benedetti E, Fiorentini C, et al. An autochthonous case of zika due to possible sexual transmission, Florence, Italy, 2014. Eurosurveillance. 2016;21(8):1–4.

29. McCarthy M. Zika virus was transmitted by sexual contact in Texas, health officials report. BMJ [Internet]. 2016;352(February):i720 . Available from: http://www.bmj.com/content/352/bmj.i720.abstract

30. Deckard DT, Chung WM, Brooks JT, Smith JC, Woldai S. Male-to-male sexual transmission of Zika virus–Texas, January 2016. Morb Mortal Wkly Rep. 2016;65(14):372–4.

31. Musso D, Roche C, Robin E, Nhan T, Teissier A, Cao-Lormeau VM. Potenial sexual transmission of Zika virus. Emerg Infect Dis. 2015;21(2):359–61.

32. Izopet J, Martin-blondel G. Correspondence Zika virus: high infectious viral load in semen, a new sexually transmitted pathogen? Lancet Infect Dis [Internet]. 2015;16(4):405. doi:10.1016/S1473-3099(16)00138-9.

33. Atkinson B, Hearn P, Afrough B, Lumley S, Carter D, Aarons EJ, et al. Detection of zika virus in semen. Emerg Infect Dis. 2016;22(5):940.

34. Oster AM, Russell K, Stryker JE, Friedman A, Kachur RE, Petersen EE, et al. Update: interim guidance for prevention of sexual transmission of Zika virus–United States, 2016. MMWR Morb Mortal Wkly Rep [Internet]. 2016;65(12):323–5.

35. UK. United Kingdom. Public Health England. Clinical advice on Zika: assessing pregnant women following travel; symptoms, transmission (includes sexual transmission), epidemiology. www.gov.uk/guidance/zika-virus. 2016. p. 1–12.

36. Prisant N, Bujan L, Benichou H, et al. Zika virus in the female genital tract. Lancet Infect Dis. 2016. Epub 11 Jul 2016. doi.:10.1016/S1473-3099(16)30193-1.

37. Brooks JT, Friedman A, Kachur R, LaFlam M, Peter PJ, et al. Update: Interim Guidance for Prevention of Sexual Transmission of Zika Virus - United States, july 2016. MMWR Morb Mortal Wkly Report, july 29, 65:745–47. 2016.

38. Davidson A, Slavinski S, Komoto K, Rakeman J, Weiss D. Suspected female-to-male sexual transmission of Zika virus – New York City, 2016. MMWR Morb Mortal Wkly Rep. 2016;65:716–7. doi:10.15585/mmwr.mm6528e2.

39. WHO. Amamentação no contexto do vírus Zika. Orientações provisórias. 25 de fevereiro de 2016.

40. Musso D, Nhan T, Robin E, Roche C. Potential for Zika virus transmission through blood transfusion demonstrated during an outbreak in French Polynesia. Euro Surveill. 2014;19(November 2013):20761.

41. Brasil P, Pereira JP, Jr., Raja Gabaglia C, et al. Zika virus infection in pregnant women in Rio de Janeiro – preliminary report. N Engl J Med. 2016. doi:10.1056/NEJMoa1602412.

42. Brito C, Cordeiro M. One year after the Zika virus outbreak in Brazil: from hypotheses to evidence. Rev Soc Bras Med Trop. 2016;49(5):537–43.

43. Cao-Lormeau V-M, Blake A, Mons S, Lastère S, Roche C, Vanhomwegen J, et al. Guillain-Barré Syndrome outbreak associated with Zika virus infection in French Polynesia: a case-control study. Lancet [Internet]. 2016;387(10027):1531–9.

44. Smith DW, Mackenzie J. Zika virus and Guillain-Barré syndrome: another viral cause to add to the list. Lancet. 2016;387(10027):1486–8.

45. Dick GWA. Zika Virus(Ii). Pathogenicity And physical properties. Trans R Soc Trop Med Hyg. 1952;46(5):521–34.

Congenital Zika Syndrome: Clinical Aspects

Vanessa van der Linden, Epitacio Leite Rolim Filho, and Ana van der Linden

Microcephaly is a condition in which a baby has a smaller head compared to other babies of the same sex and age. An infant is considered to have microcephaly when the head circumference (also known as occipitofrontal circumference) is less than a specific cutoff value compared to head circumference reference standards for boys or girls of equivalent gestational or postnatal age. Head circumference reflects intracranial volume, and it is an important measurement to monitor a child's brain growth.

Microcephaly can be caused by numerous genetic factors and also nongenetic etiologies. Nongenetic causes include congenital infections, notably the TORCH infections (toxoplasmosis,

V. van der Linden (✉)
Hospital Barão de Lucena, Recife, PE, Brazil
e-mail: vanessavdlinden@hotmail.com

E.L. Rolim Filho
Universidade Federal de Pernambuco, Recife, PE, Brazil
e-mail: filhorolim@gmail.com

A. van der Linden
IMIP—Instituto Materno Infantil Prof. Fernando Figueira,
Recife, PE, Brazil
e-mail: anavdlinden@uol.com.br

M.F.V.V. Aragão (ed.), *Zika in Focus*,
DOI 10.1007/978-3-319-53643-9_3,
© Springer International Publishing AG 2017

rubella, cytomegalovirus, and herpes), syphilis, varicella zoster, parvovirus B19, and human immunodeficiency virus (HIV) [1].

Increased rates of congenital microcephaly have been reported in settings of Zika virus transmission in Brazil which has started in late 2015 and French Polynesia from 2013 to 2015. Since the World Health Organization advised that the clusters of microcephaly and other neurological disorders and their possible association with Zika virus constituted a Public Health Emergency of International Concern, efforts have been made to describe and understand the syndrome [2].

Zika virus infection during pregnancy is a cause of microcephaly and other serious brain anomalies; however, the clinical spectrum of the effects of Zika virus infection during pregnancy is not yet known. The range of abnormalities seen and the likely causal relationship with Zika virus infection suggest the presence of a new congenital syndrome (congenital Zika syndrome) [3].

Clinical features of congenital Zika syndrome (CZS) are a consequence of direct neurological damage and severe intracranial volume loss. The main finding of CZS is microcephaly.

Microcephaly is defined when the head circumference (HC) at birth is two standard deviations below the mean for gestational age and sex. Microcephaly is considered severe when the head circumference of birth is three standard deviations below the mean for gestational age and sex [1].

A CZS has, as a main characteristic, the brain impairment, with microcephaly; however, its clinical spectrum includes newborns with normal head circumference at birth [4]. At the end of 2016 in Recife, some patients were late diagnosed with CZS; infants referred for a reference center for a CZS, after 6 months of age, presented a neurodevelopment delay with the characteristic findings in brain image described for CZS. The cases described with CZS and microcephaly at birth are, probably, the most severe spectrum of the disease, and it will be very important to follow up the cohort of the babies of pregnant women who were diagnosed with Zika infection in order to reach a better understanding of the full spectrum of this syndrome.

Severe microcephaly (more than 3 SD below the mean) observed with intrauterine Zika virus infection can be accompanied by findings consistent with fetal brain disruption sequence (FBDS) [5]. Fetal brain disruption sequence is characterized by a severe microcephaly, overlapping cranial sutures, prominent occipital bone, and redundant scalp skin, in addition to severe neurologic impairment. There is often extreme craniofacial disproportion with depression of the frontal bones and parietal bones, which can overlap [5, 6] (Fig. 1).

The FBDS phenotype is hypothesized to be a result of brain volume loss and decrease in intracranial pressure, and

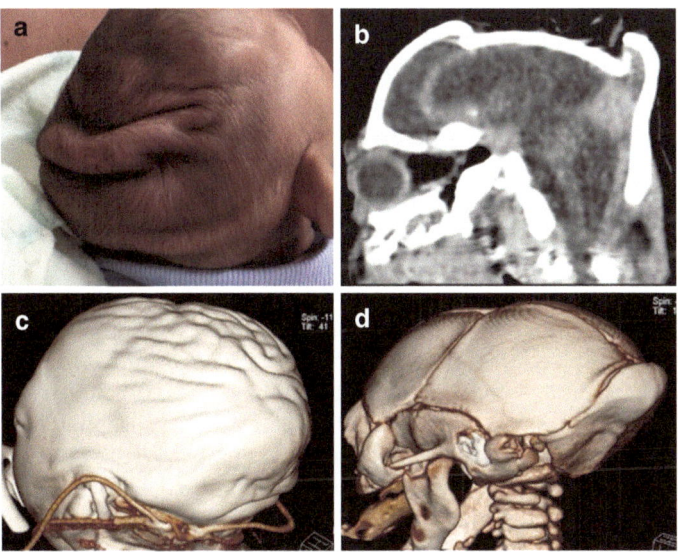

FIGURE 1 Pictures of two different patients. (**a**, **b**) Newborn pictures and CT scan from one of them, and (**c**, **d**) a CT scan with reconstruction from another; all images show the pattern of fetal brain disruption sequence characterized by a severe microcephaly, overlapping cranial sutures, prominent occipital bone, and redundant scalp skin

it is not specific to the etiologic agent. While FBDS is not unique to CZS, the phenotype was rarely reported before [6].

In addition to congenital microcephaly and craniofacial disproportion, a range of manifestations, including neurologic symptoms; limb contracture, including arthrogryposis; hearing and ocular abnormalities; and brain anomalies detected by neuroimaging, have been reported among neonates who had been exposed to Zika virus in the uterus [7–10].

The neurologic findings of severely affected patients include irritability, hyperexcitability, hypertonia, and dysphagia with feeding problems and secondary respiratory complications. Regardless of the ophthalmologic impairment, more than a half of the patients with the CZS and severe microcephaly have presented poor interaction with the environment related to cortical impairment. Motor disabilities are constantly found in patients with the most severe form of the disease, with the presence of pyramidal and extrapyramidal signs, usually associated with dystonic movement. These motor changes are probably correlated with the neuroimaging and pathological changes described, with predominant neural involvement.

van der Linden et al. described a series of infants with CZS and normal head circumference at birth documented to have poor head growth with microcephaly development after birth. Decreases in the rate of head postnatal growth, in these infants, were accompanied by significant neurologic dysfunction, including hypertonia and hemiparesis, dyskinesia/dystonia, dysphagia, epilepsy, and persistence of primitive reflexes [4]. Although these neurologic findings are consistent with previous reports of infants with congenital microcephaly who had prenatal exposure to Zika virus, infants who did not have microcephaly at birth have shown a better social interaction (i.e., they have made and held eye contact and had a social smile). However, more than 60% of infants in this series have had epilepsy, and all of them have had significant motor disabilities consistent with mixed cerebral palsy. Some of the children with normal head circumference at birth seemed to have a disproportionately small head compared to the face (craniofacial disproportion), which may suggest relatively poor brain growth [4] (Fig. 2).

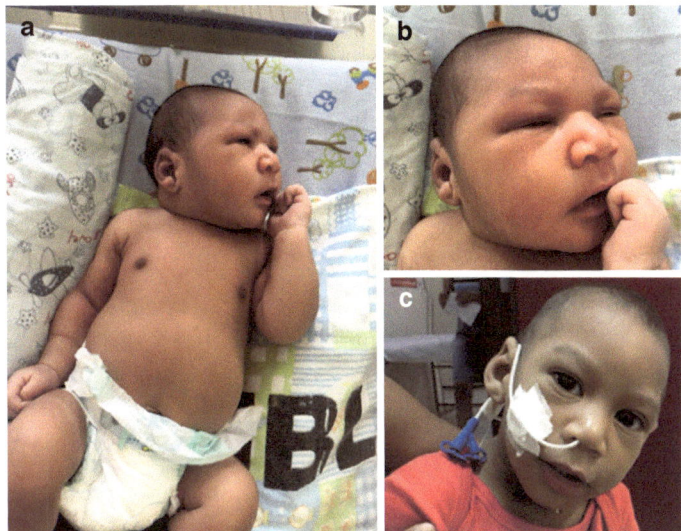

FIGURE 2 Three pictures of a patient with congenital Zika syndrome that presented normal head circumference at birth, but with craniofacial disproportion; (**a**) and (**b**) are from the newborn period and (**c**) at 1 year old. Reproduced with permission from van der Linden [4]

Epilepsy is also frequent, even in patients with normal head circumference at birth, predominating asymmetric spasms in cluster. Alves et al. described a series of 106 infants with congenital Zika syndrome who presented a high incidence of epileptic seizures before the end of the first semester of life, and the spasm was the epileptic seizure mostly observed. In these series, 40 children (38.7%) presented an epileptic seizure, classified at 43.3% of the cases as being spasms, 22.7% as generalized tonic seizures, 20.5% as partial seizures, and 4.5% as other types of seizures [11].

Unlike other congenital infections, several patients are developing hydrocephalus between 3 to 12 months, with no clinical symptoms or nonspecific symptoms. The pathophysiology of hydrocephalus in CZS is still unknown, but we believe that all patients should have a control of brain imaging between 10 to 12 months of age or in the presence of unspecific clinical worsening, such as an increase in the number of seizures and irritability or specific symptoms, with vomiting and increased head circumference. More research is needed to better understand the relationship between CZS and hydrocephalus and to define the appropriate period for reevaluation with brain imaging (Fig. 3).

Arthrogryposis is derived from the Greek words arthro (joint) and gryposis (crooked). The term arthrogryposis is often used as

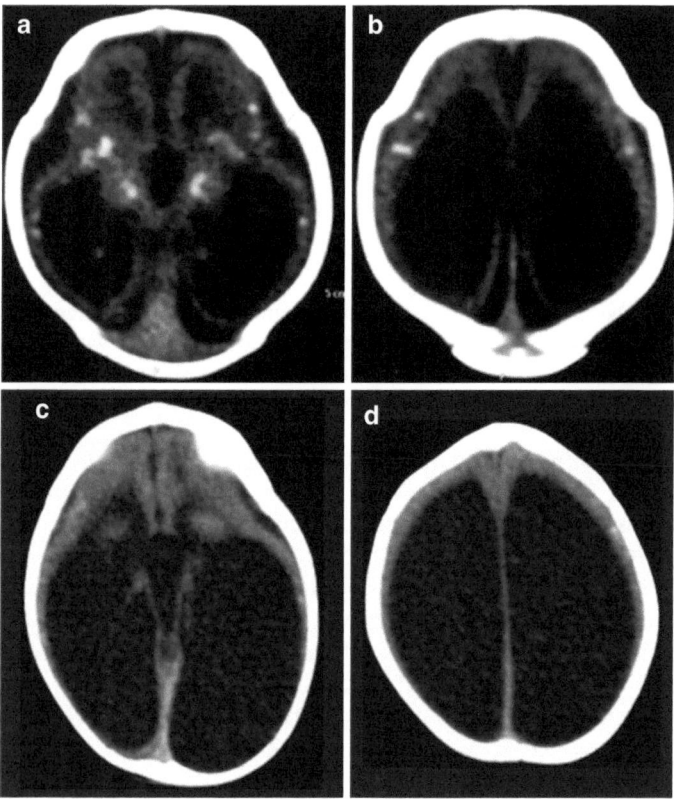

FIGURE 3 A brain CT scan of one patient that developed hydrocephaly diagnosed at 13 months of age. (**a**, **b**) Brain CT scan at the first month of age and (**c**, **d**) brain CT scan at 13 months of age; increase of the size of the ventriculus characterizing a communicating hydrocephalus was evident

shorthand to describe multiple congenital contractures that affect two or more different areas of the body. Arthrogryposis is not a specific diagnosis, but rather a clinical finding, and it is a characteristic of more than 300 different disorders [12]. Arthrogryposis can be divided into subgroups, as a way of generating a differential diagnosis, which include neurological diseases (brain, spine, or peripheral nerve), connective tissue defects (diastrophic dysplasia), muscle abnormalities (muscular dystrophies or mitochondrial abnormalities), space limitations within the uterus (oligohydramnios, fibroids, uterine malformations, or multiple pregnancy), intrauterine or fetal vascular compromise (normal development of nerves impaired or anterior horn cell death), and maternal diseases (diabetes mellitus, multiple sclerosis, myasthenia gravis, infection, drugs, or trauma) [13].

Neurologic abnormalities seem to be one of the most common causes of arthrogryposis (approximately 70–80% of cases) [13]. Developmental abnormalities affecting the forebrain (e.g., hydranencephaly, microcephaly, or forebrain neuronal migration disorders), whether due to primarily genetic factors or as a consequence of fetal central nervous system infection, are sometimes associated with arthrogryposis. In most of the cases, joint contractures are probably owing to diminished corticospinal tract activation of spinal cord motor neurons, or the underlying disease also directly injures spinal cord motor neurons, contributing to fetal hypomotility [12, 13].

By 2015, there was not a report of congenital infections associated with arthrogryposis in humans. Schuler-Faccini et al. and Oliveira Melo et al. described the association between arthrogryposis and microcephaly in newborns presumed to have been infected by congenital Zika virus, and van der Linden et al. (2016) described the main findings in a case series of arthrogryposis [7, 14, 15].

The rare arthrogrypotic joints in patients with CZS did not result from abnormalities of the joints themselves and are likely to be of neurogenic origin, with chronic involvement of central and peripheral motor neurons, leading to intrauterine fixed postures and consequently deformities (Fig. 4). Electromyography findings suggest chronic involvement of

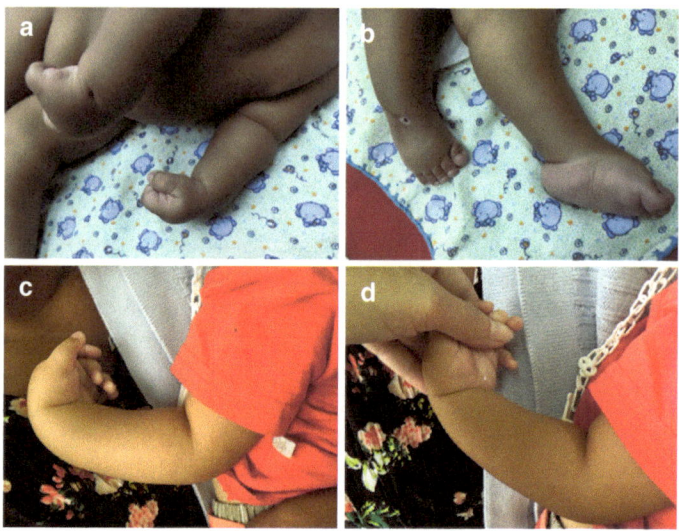

FIGURE 4 Pictures of two different patients with CZS and arthrogryposis, (**a**) and (**b**) from one patient and (**c**) and (**d**) from another patient. The pattern of malformation of the hand in pictures (**a**), (**c**), and (**d**) are common findings between patients with CZS and arthrogryposis. Figs. a and b reproduced with permission from de Fátima Vasco Aragão et al. [16]

peripheral motor neurons. In severely weak muscles, the activation of the motor units was severely reduced; it was suggesting a reduced central drive and involvement of central motor neurons. The pattern of peripheral denervation seems to correspond to the pattern of central involvement, which could suggest a component of transsynaptic degeneration. The spine MRI of the patients with CZS and arthrogryposis shows apparent thinning of the spinal cord, and reduced ventral roots of medullary cone corroborate the findings of electromyography [7]. Some patients presented abnormal posture of the limbs, with similar pattern of distribution of the patients with arthrogryposis but without fixed deformity, and the same electromyography findings suggest chronic involvement of peripheral motor neurons (Fig. 5).

Studies in experimental models have implicated neural progenitor cells as a primary Zika virus target; however,

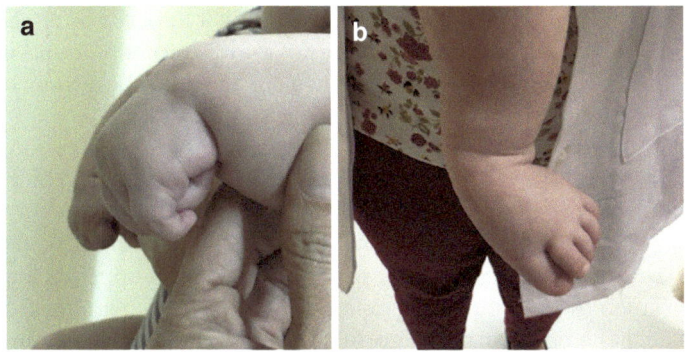

Figure 5 Pictures of one patient with abnormal posture of the hand (**a**)and foot (**b**)without fixed deformity; electromyography findings suggest chronic involvement of peripheral motor neurons

immature neurons were also infected to a lesser extent [17]. On microscopic examination of a Zika virus-infected fetal brain, postmigratory neurons—primarily intermediately differentiated—were apoptotic [18]. Those findings supported direct neural cell injury by Zika virus and suggested disruption of existing immature neurons, as well as decreased proliferation and impaired migration due to progenitor cells loss [17]. Zika virus is a neurotropic virus that particularly targets neural progenitor cells. Murine and human placental studies support the hypothesis that maternal infection leads to placental infection and injury, followed by transmission of the virus to the fetal brain, where it kills neuronal progenitor cells and disrupts neuronal proliferation, migration, and differentiation, which slows brain growth and reduces viability of neural cells [19].

Neurophysiological observations in patients with CZS and arthrogryposis and the literature finding suggest a viral tropism for the neurons or neural progenitor cells, with involvement of peripheral motor neurons and central motor neurons.

The orthopedic abnormalities found on patients with CZS can be divided into congenital (primary), which is presented on the child's birth, and secondary, which appears after the child's birth. The primary deformities are normally found in the arthrogryposis carriers which, in this specific group of patients, probably have neurogenic origin.

The most probable cause of the secondary deformities is the abnormal muscle movements, caused by the primary alterations of the central and peripheral (or both) nervous system.

The data described below were based on findings of the orthopedic tests: simple X-ray of the axial and appendicular skeleton and ultrasonography of 157 patients with microcephaly between the age of 0 and 15 months, who were treated by the Association for Assistance of Disabled Children (AACD) in Recife, Brazil.

Among the 157 patients who were observed, 102 (64.96%) patients presented some orthopedic deformity primary or secondary. In total, there were 250 musculoskeletal abnormalities: 44 (17.6%) in the upper limbs (graphic 1), 150 (62%) in the lower limbs (graphic 2), and 56 (20.4%) in the trunk and abdomen (graphic 3). Forty-six (18.4%) were primary orthopedic congenital alterations, in this case, normally presented in those patients who were diagnosed with arthrogryposes.

The most frequent congenital deformities observed in those patients are the congenital teratological dislocation of the hips followed by the clubfoot deformity and followed by the camptodactyly of the fingers and subluxation or dislocation of the knees. These deformities are normally found in patients who were diagnosed with arthrogryposes (Fig. 6). So, they coexist in the same patient, whereas the isolated congenital clubfoot, evolutionary hip dysplasia, and other isolated congenital deformities, at first, seemed not to have a greater incidence if compared to the population in general.

The simple X-ray images of the axial and appendicular skeleton of these patients did not show any dysplastic alteration of the long bones and spine. Hemimelic and transverse limb deformities, defects of duplication, and other bone dysostoses were also not found in any of the 157 patients who were evaluated.

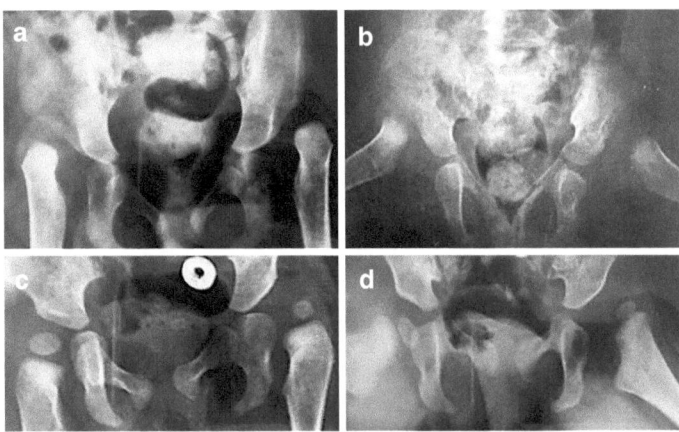

FIGURE 6 Anteroposterior (**a**) and Froger (**b**) radiographs showing features compatible with dislocation of hips. Anteroposterior (**c**) and Froger (**d**) radiographs showing features compatible with hip development dysplasia. Dysplastic changes are more severe in teratological dislocations that are generally irreducible to clinical maneuvers

As it was described above, the developmental dysplasia of the hip (DDH) did not seem to have a larger incidence in these patients if compared with the population in general. Except for the congenital dislocation observed in patients with arthrogryposis, therefore a type of teratological dislocation, the hip ultrasonographic evaluation, by the GRAF method, of 87 infants, from 0 to 2 months with CZS, performed by us in the AACD-Recife, showed that only two patients had presented alterations related to evolutionary dysplasia of the hips; all other dislocations of the hips were teratological, associated with arthrogryposis.

Considering that these two children are female and daughters of mothers in first pregnancy, factors which are known to be related to a greater risk of DDH (the developmental dysplasia of the hip), in the general population, it may be suggested that DDH is not a common finding in this patients but probably secondary to the primary neurological impairment.

References

1. World Health Organization. Screening, assessment and management of neonates and infants with complications associated with Zika virus exposure in utero. Updated 30 Aug 2016, WHO reference number: WHO/ZIKV/MOC/16.3 Rev.3.
2. World Health Organization. WHO statement on the first meeting of the International Health Regulations (2005) (IHR 2005) Emergency Committee on Zika virus and observed increase in neurological disorders and neonatal malformations. http://www.who.int/mediacentre/news/statements/2016/1st-emergency-committee-zika/en/. Accessed 01 Feb 2016.
3. Costello A, Dua T, Duran P, et al. Defining the syndrome associated with congenital Zika virus infection. Bull World Health Organ. 2016;94(6):406–406A.
4. van der Linden V, Pessoa A, Dobyns W, et al. Description of 13 infants born during October 2015–January 2016 with congenital Zika virus infection without microcephaly at birth—Brazil. http://www.cdc.gov/mmwr/volumes/65/wr/mm6547e2.htm?s_cid=mm6547e2_e. Accessed 22 Nov 2016.
5. Moore CA, Staples EJ, Dobyns WB, et al. Characterizing the pattern of anomalies in congenital Zika syndrome for pediatric clinicians. JAMA Pediatr. Published online 3 Nov 2016.
6. Russell LJ, Weaver DD, Bull MJ, Weinbaum M. In utero brain destruction resulting in collapse of the fetal skull, microcephaly, scalp rugae, and neurologic impairment: the fetal brain disruption sequence. Am J Med Genet. 1984;17(2):50.
7. van der Linden V, Filho ER, Lins OG, van der Linden A, Aragão Mde F, Brainer-Lima AM, et al. Congenital Zika syndrome with arthrogryposis: retrospective case series study. BMJ. 2016;354:i3899.
8. Leal MC, Muniz LF, Ferreira TS, et al. Hearing loss in infants with microcephaly and evidence of congenital Zika virus infection—Brazil, November 2015–May 2016. MMWR Morb Mortal Wkly Rep. 2016.
9. Ventura CV, Maia M, Ventura BV, et al. Ophthalmological findings in infants with microcephaly and presumable intra-uterus Zika virus infection. Arq Bras Oftalmol. 2016;79:1.
10. de Fátima Vasco Aragão M, van der Linden V, Brainer-Lima AM, Coeli RR, Rocha MA, Sobral da Silva P, et al. Clinical features and neuroimaging (CT and MRI) findings in presumed Zika virus related congenital infection and microcephaly: retrospective case series study. BMJ. 2016;353:i1901.

11. Alves LV, Cruz DDCS, van der Linden A, et al. Epileptic seizures in children with congenital Zika virus syndrome. Rev Bras Saúde Matern Infant Recife. 2016;16(1):S27–31.
12. Bamshad M, Heest AE, Pleasure D. Arthrogryposis: a review and update. J Bone Joint Surg Am. 2009;91(Suppl 4):40–6.
13. Kalampokas E, Kalampokas T, Sofoudis C, Deligeoroglou E, Botsis D. Diagnosing arthrogryposis multiplex congenita: a review. ISRN Obstet Gynecol. 2012;2012:264918.
14. Schuler-Faccini L, Ribeiro EM, Feitosa IM, Horovitz DD, Cavalcanti DP, Pessoa A, et al. Possible association between Zika virus infection and microcephaly—Brazil, 2015. MMWR Morb Mortal Wkly Rep. 2016;65(3):59–62.
15. Oliveira Melo AS, Malinger G, Ximenes R, Szejnfeld PO, Alves Sampaio S, Bispo de Filippis AM. Zika virus intrauterine infection causes fetal brain abnormality and microcephaly: tip of the iceberg? Ultrasound Obstet Gynecol. 2016;47(1):6–7.
16. de Fátima Vasco Aragão M, Brainer-Lima AM, Holanda AC, van der Linden V, Aragão LV, Silva Júnior MLM, Sarteschi C, Petribu NCL, van der Linden A, Valença MM. (IN PRESS) Spectrum of spinal cord, spinal roots and brain MRI abnormalities in congenital Zika syndrome with and without arthrogryposis. AJNR. 2017.
17. Li C, Xu D, Ye Q, et al. Zika virus disrupts neural progenitor development and leads to microcephaly in mice. Cell Stem Cell. 2016;19:1–7. doi:10.1016/j.stem.2016.04.017.
18. Mlakar J, Korva M, Tul N, Popović M, Poljšak-Prijatelj M, Mraz J, et al. Zika virus associated with microcephaly. N Engl J Med. 2016. doi:10.1056/NEJMoa1600651.
19. Garcez PP, Loiola EC, Madeiro da Costa R, et al. Zika virus impairs growth in human neurospheres and brain organoids. Science. 2016;352:816.

Ocular Findings in Children with Congenital Zika Syndrome

Camila V. Ventura, Natalia C. Dias, Audina M. Berrocal, and Liana O. Ventura

C.V. Ventura, M.D. (✉)
Altino Ventura Foundation, Recife, PE, Brazil

Department of Ophthalmology, Pernambuco's Eye Hospital (HOPE), Recife, PE, Brazil

Department of Ophthalmology and Visual Sciences, Paulista School of Medicine, Federal University of São Paulo, São Paulo, Brazil

Department of Ophthalmology, Bascom Palmer Eye Institute, University of Miami Miller School of Medicine, Miami, FL, USA
e-mail: camilaventuramd@gmail.com

N.C. Dias, M.D.
Altino Ventura Foundation, Recife, PE, Brazil
e-mail: nataliadecarvalhodias@gmail.com

A.M. Berrocal, M.D.
Department of Ophthalmology, Bascom Palmer Eye Institute, University of Miami Miller School of Medicine, Miami, FL, USA
e-mail: aberrocal@med.miami.edu

L.O. Ventura, M.D., Ph.D.
Altino Ventura Foundation, Recife, PE, Brazil

Department of Ophthalmology, Pernambuco's Eye Hospital (HOPE), Recife, PE, Brazil
e-mail: lianaventuramd@gmail.com

M.F.V.V. Aragão (ed.), *Zika in Focus*,
DOI 10.1007/978-3-319-53643-9_4,
© Springer International Publishing AG 2017

Introduction

The zones of active Zika virus infection are expanding globally with the northeastern part of Brazil being a major epicenter of the virus [1, 2]. When Zika virus first stroke Brazil in May 2015, the population was by far unprepared for the subsequent implications of this new disease to the Americas [1, 3].

At that time, Zika virus was not considered a threat, being perceived by many as a harmless virus that caused solely mild symptoms (headache, rash, arthralgia, and conjunctivitis) to only 20% of the infected people [3, 4]. This flavivirus, considered endemic in Africa and Asia for more than 50 years, had never been associated with any congenital malformations nor with neurological abnormalities [4]. However, in October 2015, several months after the first autochthonous cases of Zika virus were reported in the northeast of Brazil, an increased number of newborns with microcephaly were observed in the same region [1, 3].

Despite being the primary and major findings identified in babies exposed to Zika virus intra-uterus, microcephaly and other neurological abnormalities are now known to compose only one of the pillars of this new entity. In addition to the dramatic neurological alterations seen in these babies, Zika virus can cause other systemic findings including ophthalmological, audiological, and skeletal malformations [5–12]. These systemic abnormalities comprise what is now called the congenital Zika syndrome (CZS) [5]. In this chapter, the ophthalmological aspects of CZS will be addressed.

The History Behind the Discovery

Before January 2016, it was thought that Zika virus caused neurological findings alone [1, 3]. However, after Ventura et al. reported the first three cases of infants with microcephaly that presented retinal findings, the scope of the disease expanded [9]. This report described the retinal alterations that infants with CZS can present, which consists of basically pigment mottling and/or chorioretinal atrophy, but have a different appearance compared to other well-known congenital infections (Fig. 1) [9].

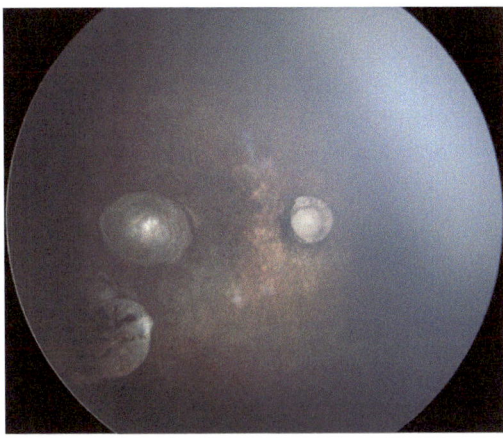

FIGURE 1 Right eye fundus photography showing optic nerve pallor and hypoplasia, two well-delineated chorioretinal scars located in the macula and close to the inferior arcade, pigment mottling in the macular region, and vascular attenuation

Subsequent publications written by the same group and by de Paula Freitas et al. reported additional cases that had not only the retinal findings previously described but also optic nerve abnormalities [11, 12]. The ocular findings described by de Paula Freitas et al. were similar to the ones described by Ventura et al. and were identified in infants from another northeastern state in Brazil called Bahia [11, 12]. In fact, De Paula Freitas' study was remarkably important to endorse what was primarily identified in babies born in Pernambuco state, the epicenter of Zika infection in Brazil.

However, in all of these studies, due to the unavailability of specific serology for Zika virus, cases were given the diagnosis of probable congenital Zika virus infection [9, 11, 12]. In truth, it was only in the late May 2016 that the first publication addressing the ocular findings in these babies included specific serology for Zika virus [13]. In this study, Zika virus testing was performed in 24/40 (60%) of infants with microcephaly using IgM antibody-capture ELISA, and 100% of those tested were positive for Zika virus exposure in the cerebrospinal fluid (CSF) [13]. Despite the fact that the current serology available is still not ideal to test Zika virus exposure due to its high cross-reactivity with other flaviviruses, it was enough to give scientists enough support to state that the vertical transmission of Zika virus is not only related to microcephaly but also to all other systemic findings, including the ocular findings [1, 3, 4, 7, 8, 13].

Pregnancy History and Prevalence of Ocular Findings

In the study conducted in Salvador, 79.3% of the mothers of babies with probable CZS referred rash, fever, arthralgia, headache, itchiness, and malaise during pregnancy, from which 62% occurred during the first trimester [12]. In the study conducted in Pernambuco, similar symptoms were reported by 67.5% of the mothers; and from the mothers of babies with ocular findings, 71.4% referred symptoms in the first trimester [13]. In this latter study, researchers were able to prove the relationship between the ocular findings and the time of infection during pregnancy; the first trimester was identified as risk factor for ocular findings in infants with CZS [13]. One interesting fact was that in neither of these studies, mothers reported conjunctivitis or other ocular symptoms when infected during pregnancy as other studies from the Orient have pointed out [4, 12, 13].

The estimated prevalence of ocular findings in both studies led by de Paula Freitas (Salvador) and Ventura (Pernambuco) was 34.5% and 55%, respectively [12, 13]. Moreover, Paula de Freitas et al. reported bilateral ocular findings in 70% of the cases and Ventura et al. in 68.2% [12, 13].

The Ocular Findings

When analyzing all previous studies that have addressed the ocular findings in babies with CZS, it is possible to infer that the retina and optic nerve are the main structures affected [9, 11–14]. The most common retinal findings include focal pigmentary mottling and chorioretinal atrophy, which are most commonly observed in the macular region [9, 11–14]. However, de Paula Freitas et al. described a case where the chorioretinal atrophy was located nasally [12].

These chorioretinal scars are typically well defined with a hyperpigmented rim demarcating the atrophic region that can vary in shape (ovoid, circular, lobulated, and colobomatous-like), size (from a small and discrete lesion, all the way to an extensive lesion occupying an entire hemiretina), and quantity (one or multiple) [9, 11–14] (Figs. 1 and 2).

The focal pigment mottling in affected eyes has been described as clusters of pigment (fine or gross) in the macular region [9, 11–14] (Fig. 3). The optic disk findings reported so far include optic disk hypoplasia, pallor, and/or increased cup-to-disk ratio [11–14].

On May 2016, Miranda et al. expanded the ocular spectrum of CZS when he reported retinal hemorrhages and peripheral vasculature abnormalities in the eyes of babies infected with Zika virus [14].

Although posterior segment findings are the most prevalent, cataracts, intraocular calcifications, and structural ocular

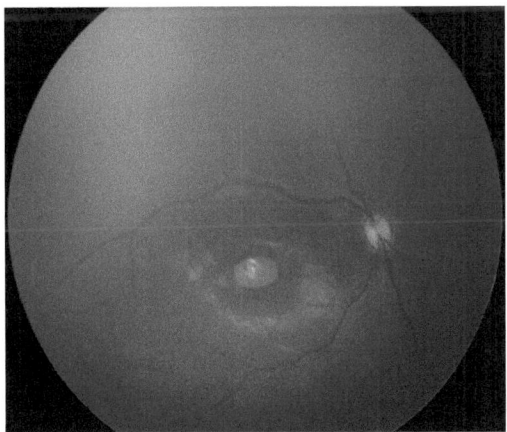

FIGURE 2 Right eye fundus photography showing optic nerve pallor and a well-delineated chorioretinal scar in the macula

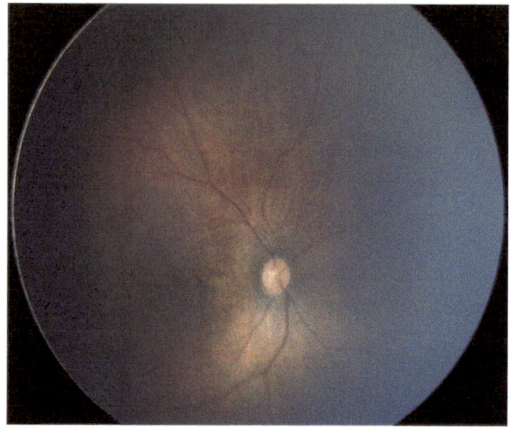

FIGURE 3 Right eye fundus photography showing focal pigment mottling in the macula

abnormalities including microphthalmia, iris coloboma, and lens subluxation have also been reported in infants with CZS [12, 15, 16].

Pathophysiology of Ocular Lesions

Laboratory experiments have been using animal models to address the pathophysiology of CZS [17–20]. Mysorekar et al. have shown that Zika virus is transmitted to fetuses via placenta and that this transmission may cause a broad spectrum of systemic abnormalities [17]. Other studies have been able to demonstrate that Zika virus attacks cortical progenitor cells in the brain, causing cell death and impairing neurodevelopment [18–20].

Although it is expected that a similar mechanism is happening to the retina of these infected babies, the specific pathophysiology of the ocular findings remains unknown. Ventura et al. have hypothesized that the retinal findings are probably directly related to the virus or to virus toxin that leads to an inflammatory reaction in the neurosensory retina [11]. However, other unknown factors such as the amount of viral circulation and the immunologic response of mother and fetus may play an important role on the abnormalities observed in the newborns [11].

Risk Factors Associated with Ocular Findings

Ventura et al. were able to isolate the main risk factors associated with the ocular findings in CZS [13]. These risk factors include the severity of microcephaly at birth and the trimester when the infection occurred. In other words, babies that were exposed to Zika virus in the first trimester of pregnancy and those born with severe microcephaly have a greater chance of presenting ocular findings at birth [13].

Ocular Findings in the Absence of Microcephaly

An important scientific paradigm shift occurred when Ventura et al. reported the first case of CZS that did not present microcephaly at birth yet presented other neurological findings and a chorioretinal scar in the left eye [21].

In the beginning, scientists thought that all babies affected by Zika virus intra-uterus would necessarily present microcephaly at birth. For months, microcephaly was a required criterion for further investigations in these babies, including serological and clinical evaluations [6, 7, 9, 11–14]. However, after this first report, it was possible to change the screening criteria in babies exposed to Zika virus during pregnancy, and this criterion became obsolete [18].

Just recently, Van der Linden et al. described another similar case of an infant with CZS and normal head circumference at birth that presented other brain findings, chorioretinal scar, and arthrogryposis [8].

The Centers of Diseases Control (CDC) Recommendations

The current Centers of Diseases Control (CDC) guidelines recommend that initial clinical evaluation should be done in all infants with laboratory evidence of Zika virus infection with or without abnormalities consistent with CZS, as well as babies with abnormal clinical or neuroimaging findings at birth [22]. The initial ocular screening should be performed before hospital discharge or within 1 month and should include fundus evaluation. It is also recommended to repeat the ophthalmological evaluation at 3 months of age for those babies with the diagnosis of CZS [22].

Ocular Findings in Adults During Acute Infection

In adults, the usual symptoms of acute Zika virus infection included headache, rash, arthralgia, and conjunctivitis [4]. However, among all interesting facts revolving the Brazilian outbreak, conjunctivitis has not been reported by the mothers infected during their pregnancy, as described in the Micronesia outbreak [4, 11–13]. In fact, what have been reported are acute uveitis and maculopathy [23–25].

Furtado et al. and Fontes et al. recently described two intriguing cases of acute bilateral hypertensive iridocyclitis in adults during the acute phase of the disease that have eased with the viremia descent [23, 24]. In addition, Parke et al. have published a case of a unilateral acute maculopathy in a male adult infected with Zika that also resolved within 6 weeks [25].

Conclusion

Although Zika virus infection in adults is often mild or not recognized, exposure to the virus during pregnancy can presumably lead to devastating effects on the developing fetus. The eye is dramatically affected in CZS, and the main structures affected in the eye are the retina and optic nerve. However, more studies are necessary to address the visual outcome and the long-term consequences of CZS in these infants.

References

1. World Health Organization (WHO). Epidemiological alert: neurological syndrome, congenital malformations, and Zika virus infection. Implications for public health in the Americas. 2015. http://www.paho.org/hq/index.php?option=com_docman&task=doc_view&Itemid=270&gid=32405&lang=en. Accessed 27 Jan 2015.
2. Centers for Disease Control and Prevention (CDC). All countries and territories with active Zika virus transmission. 2016. Last update 5 Oct 2016. Accessed 29 Oct 2016.
3. Brasil-Ministério da Saúde. Protocolo de vigilância e resposta à ocorrência de microcefalia relacionada à infecção pelo vírus Zika. 2016. http://portalsaudesaudegovbr/images/pdf/2015/dezembro/09/Microcefalia---Protocolo-de-vigil--ncia-e-resposta--vers--o-1----09dez2015-8hpdf Acessed 26 Jan 2016.
4. Petersen LR, Jamieson DJ, Powers AM, Honein MA. Zika virus. N Engl J Med. 2016;374(16):1552–63.
5. Miranda-Filho Dde B, Martelli CM, Ximenes RA, et al. Initial description of the presumed congenital Zika syndrome. Am J Public Health. 2016;106(4):598–600.
6. De Fátima Vasco Aragão M, van der Linden V, Brainer-Lima AM, et al. Clinical features and neuroimaging (CT and MRI) findings in presumed Zika virus related congenital infection and microcephaly: retrospective case series study. BMJ. 2016;353:i901.
7. Leal MC, Muniz LF, Ferreira TS, et al. Hearing loss in infants with microcephaly and evidence of congenital virus infection—Brazil, November 2015—May 2016. MMWR. 2016;65:1–4.
8. Van der linden V, Rolim Filho EL, Lins OG, et al. Congenital Zika syndrome with arthrogryposis: retrospective series study. BMJ. 2016;354:i3899.
9. Ventura CV, Maia M, Bravo-Filho V, Gois AL, Belfort Jr R. Zika virus in Brazil and macular atrophy in a child with microcephaly. Lancet. 2016;387(10015):228.

10. Mets MB, Chhabra MS. Eye manifestations of intrauterine infections and their impact on childhood blindness. Surv Ophthalmol. 2008;53(2):95–111.

11. Ventura CV, Maia M, Ventura BV, et al. Ophthalmologic assessment of ten infants with microcephaly and presumable intrauterus zika virus infection. Arq Bras Oftalmol. 2016;79(1):1–3.

12. de Paula FB, de Oliveira Dias JR, Prazeres J, et al. Ocular findings in infants with microcephaly associated with presumed zika virus congenital infection in Salvador, Brazil. JAMA Ophthalmol. 2016;134(5):529–35.

13. Ventura CV, Maia M, Travassos SB, et al. Risk factors associated with the ophthalmoscopic findings identified in infants with presumed Zika virus congenital infection. JAMA Ophthalmol. 2016;134(8):912–8.

14. Miranda HA, Costa MC, Frazão MA, Simão N, Franchischini S, Moshfeghi DM. Expanded spectrum of congenital ocular findings in microcephaly with presumed Zika infection. Ophthalmology. 2016. pii:S0161-6420(16)30277-30279.

15. Oliveira Melo AS, Malinger G, Ximenes R, Szejnfeld PO, Alves Sampaio S, Bispo de Filippis AM. Zika virus intrauterine infection causes fetal brain abnormality and microcephaly: tip of the iceberg? Ultrasound Obstet Gynecol. 2016;47(1):6–7.

16. Calvet G, Aguiar RS, Melo AS, et al. Detection and sequencing of Zika virus from amniotic fluid of fetuses with microcephaly in Brazil: a case study. Lancet Infect Dis. 2016;16(6):653–60.

17. Mysorekar IU, Diamond MS. Modeling Zika virus infection in pregnancy. N Engl J Med. 2016;375(5):481–4.

18. Cugola FR, Fernandes IR, Russo FB, et al. The Brazilian Zika virus strain causes birth defects in experimental models. Nature. 2016;534(7606):267–71.

19. Li C, Xu D, Ye Q, et al. Zika virus disrupts neural progenitor development and leads to microcephaly in mice. Cell Stem Cell. 2016;19(1):120–6.

20. Wu K, Zuo G, Li X, et al. Vertical transmission of Zika virus targeting the radial glial cells affects cortex development of offspring mice. Cell Res. 2016;26(6):645–54.

21. Ventura CV, Maia M, Dias N, Ventura LO, Belfort Jr R. Zika: neurological and ocular findings in infant without microcephaly. Lancet. 2016;387(10037):2502.

22. Russell K, Oliver SE, Lewis L, et al. Update: interim guidance for the evaluation and management of infants with possible

congenital Zika virus infection—United States, August 2016. MMWR Morb Mortal Wkly Rep. 2016;65:870–8.

23. Furtado JM, Espósito DL, Klein TM, Teixeira-Pinto T. Uveitis associated with Zika virus infection. N Engl J Med. 2016;375:394–6.

24. Fontes BM. Zika virus-related hypertensive iridocyclitis. Arq Bras Oftalmol. 2016;79(1):63.

25. Parke 3rd DW, Almeida DR, Albini TA, Ventura CV, Berrocal AM, Mittra RA. Serologically congirmed Zika-related unilateral acute maculopathy in an adult. Ophthalmology. 2016;123(11):2432–3.

Zika Virus: Laboratory Diagnosis

Marli Tenório Cordeiro

Zika virus, dengue virus (DENV), and chikungunya virus (CHIKV) share similar symptoms of infection; thus, laboratory diagnosis is an essential tool for disease confirmation, patient management, and public health prevention measures. Among flaviviruses, Zika virus and DENV are closely related, so serologic diagnosis of Zika is challenging, mainly in patients where Zika was the secondary flavivirus infection [1].

The current recommendations for routine diagnosis of Zika infection are the detection of the viral nucleic acid by real-time reverse transcription-polymerase chain reaction (rRT-PCR) or by the conventional RT-PCR, detection of Zika virus-specific IgM antibodies by the enzyme-linked immunosorbent assay (ELISA), and the plaque reduction neutralization test (PRNT) for confirmation, when necessary, of positive IgM results and exclusion of other flaviviruses [2, 3].

The presence of Zika virus within the first 5–7 days of infection (acute phase of disease) may be investigated in blood, urine, cerebrospinal fluid (CSF), and other fluids, by molecular methods and also by virus isolation in cell cultures.

M.T. Cordeiro, Ph.D.
Fiocruz, Recife, PE, Brazil
e-mail: marli@cpqam.fiocruz.br

M.F.V.V. Aragão (ed.), *Zika in Focus*,
DOI 10.1007/978-3-319-53643-9_5,
© Springer International Publishing AG 2017

However, in urine, the virus may be detected over a longer period, viz., 15–20 days from the onset of symptoms.

The ELISA is used for the qualitative detection of Zika virus-IgM antibodies in serum or CSF obtained from persons suspected of Zika infection. Detection of Zika virus-specific IgM antibodies may confirm the diagnosis, provided that the serologic test was conducted and interpreted with criteria, since cross-reaction among the flaviviruses may occur. Zika virus-specific IgM antibodies are formed during the first week of disease and may be detected from day 7 on. Two serum samples should be analyzed, one sample collected during the acute phase of illness and the other during the convalescence (14–21 days). Samples should be tested, simultaneously, for antibodies to Zika virus, DENV, and other flavivirus endemic to the region where the exposure occurred to evaluate false-positive result that may be caused by cross-reactivity ([1, 2]; Fig. 1).

If either the Zika virus- or DENV-IgM antibody testing yields positive or inconclusive results, PRNT against these

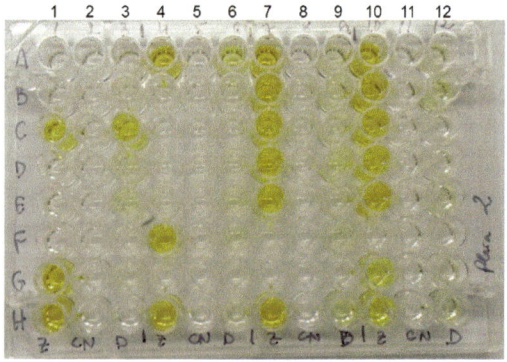

FIGURE I Zika virus- and DENV IgM-capture ELISA. *Z* Zika virus antigen, *CN* antigen normal control, *D* DENV antigen (mixture of all serotypes), *C1* Zika virus-positive control, *C3* DENV-positive control, *B1* Zika virus-negative control, *B3* DENV-negative control, in *yellow* are IgM-positive samples; in *color* are negative IgM samples

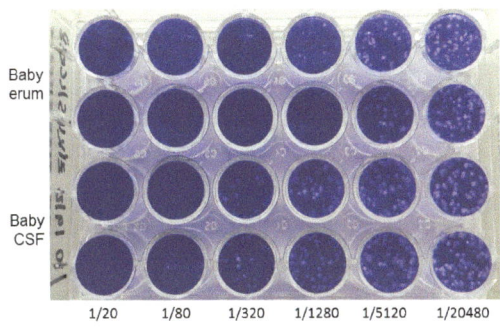

FIGURE 2 Plaque reduction neutralization test (PRNT50) for Zika virus

viruses should be performed. Particularly in pregnant women and in severe cases, it is strongly recommended that Zika virus-positive IgM be confirmed by PRNT, performed against other flaviviruses endemic to the region, to avoid false-positive ELISA results ([3]; Fig. 2).

Samples were diluted 1/20 to 1/20.480; Zika-neutralizing antibodies were detected in the baby serum sample until 1/1280 and 1/80 in the CSF sample.

It is known that DENV-IgM antibodies may be detected, approximately 90 days after the onset of symptoms. However, studies have shown that patients with West Nile virus neurologic disease had detectable IgM antibodies more than one year after illness onset [4]. So far, it is not known how long Zika virus-IgM antibodies remain detectable after infection. Since IgM does not normally cross either the placenta or the blood-brain barrier, detection of Zika virus-specific IgM in the blood of neonate suspected of congenital Zika infection, in the CSF of a neonate with brain abnormalities or in the CSF of a person with neurologic manifestations, represents a valuable diagnostic finding that confirm the Zika virus infection [5, 6].

Thus, an accurate Zika laboratory diagnosis requires combining serologic data to rRT-PCR testing, as well as the clinical and epidemiological criteria for suspected Zika disease.

References

1. Lanciotti RS, Kosoy OL, Laven JJ, Velez JO, Lambert AJ, Johnson AJ, et al. Genetic and serologic properties of Zika virus associated with an epidemic, Yap State, Micronesia, 2007. Emerg Infect Dis. 2008;14:1232–9.
2. Musso D, Gubler DJ. Zika virus. Clin Microbiol Rev. 2016;29:487–524.
3. Staples JE, Dziuban EJ, Fischer M, Cragan JD, Rasmussen SA, Cannon MJ, et al. Interim guidelines for the evaluation and testing of infants with possible congenital Zika virus infection—United States, 2016. MMWR Morb Mortal Wkly Rep. 2016;65:63–7.
4. Roehring JT, Nash D, Maldin B, et al. Persistence of virus-reactive serum immunoglobulin M antibody in confirmed West Nile encephalitis cases. Emerg Infect Dis. 2003;9:376–9.
5. Cordeiro MT, Pena LJ, Brito CA, Gil LH, Marques ET. Positive IgM for Zika virus in the cerebrospinal fluid of 30 neonates with microcephaly in Brazil. Lancet. 2016b;387(10030):1811–2. doi:10.1016/S0140-6736(16)30253-7.
6. Cordeiro MT, Brito CAA, Pena LJ, Castanha PMS, Gil LHV, Lopes KGS, Dhalia R, Meneses JA, Ishigami AC, Mello LM, Alencar LXE, Guarines KM, Rodrigues LC, Marques ETA. Results of a Zika Virus (ZIKV) Immunoglobulin M–specific diagnostic assay are highly correlated with detection of neutralizing anti-ZIKV antibodies in neonates with congenital disease. J Infect Dis. 2016a;214:1897–904.

Neuroimaging Findings of Congenital Zika Syndrome

Maria de Fátima Viana Vasco Aragão,
Alessandra Mertens Brainer-Lima, Arthur Cesário de Holanda,
and Natacha Calheiros de Lima Petribu

The imaging and pathological findings in congenital Zika syndrome include moderate and severe microcephaly, in as many as 75% of the cases [1], ophthalmological and auditory abnormalities [2, 3], and arthrogryposis [4]. However, these are not the only abnormalities that can be found, and the complete disease spectrum is still unknown. Some babies are born with head circumference at the lower limit of the normal range, but later progress to microcephaly. In the authors' experience, they correspond to about 10% of microcephaly cases confirmed with positive IgM CSF (unpublished data).

M.F.V.V. Aragão, M.D., Ph.D. (✉)
Mauricio de Nassau University, Recife, PE, Brazil

Centro Diagnóstico Multimagem, Recife, PE, Brazil
e-mail: fatima.vascoaragao@gmail.com

A.M. Brainer-Lima, M.D., M.Sc.
PROCAPE, Universidade de Pernambuco, Recife, PE, Brazil
e-mail: mertensbrainer@yahoo.com.br

A.C. Holanda
Universidade Federal de Pernambuco, Recife, PE, Brazil
e-mail: arthur.c.holanda@gmail.com

N.C. de Lima Petribu, M.D., M.Sc.
Hospital Barão de Lucena, Recife, PE, Brazil
e-mail: natachacalheiros@yahoo.com.br

M.F.V.V. Aragão (ed.), *Zika in Focus*,
DOI 10.1007/978-3-319-53643-9_6,
© Springer International Publishing AG 2017

However, the clinical spectrum of the effects of Zika virus infection during pregnancy is not yet known.

The existence of children without microcephaly with alterations of the image raises questions of great importance for public health. It is probable that we have been seeing only the "tip of the iceberg" corresponding to the more severe brain damage cases with microcephaly [5]. We do not know the real size of the submerged part of this congenital Zika syndrome "iceberg," with its minor changes and without microcephaly, which will probably lead to future problems, such as epilepsy and cognitive and motor impairment.

On neurological examination, the children with congenital Zika syndrome and microcephaly can present hypertonia, hypotonia, spasticity, hyperreflexia, irritability, and seizures [6]. These infants may frequently show ocular lesions such as atrophy and chorioretinal scars, pigmentary changes, optic disc pallor, optic nerve hypoplasia, increased pallor and optic disc excavation, hemorrhagic retinopathy, and abnormal retinal vasculature [3, 7].

Microcephaly is the most prominent and recognized manifestation of the congenital Zika syndrome. It is based on the measurement of the skull at least 24 h after birth and up to the first week of life (6 days and 23 h of life) [8]. According to the World Health Organization (WHO), a head circumference (HC) of below two standard deviations from the specific mean for sex and gestational age is considered as moderate microcephaly, while a HC below three standard deviations is considered as severe microcephaly [8].

The diagnosis of congenital syndrome by the Zika virus is currently based on clinical and radiological findings, associated with laboratory exclusion of other congenital infections or hereditary conditions (i.e., TORCHs and pseudo-TORCHs). Newborns with microcephaly should undergo an imaging study to evaluate the brain damage because the detection of calcifications is a strong hallmark of congenital infection and is usually the first diagnostic consideration suggested when it is identified [9].

Considering that this epidemic is a public health problem, if the fontanel is large enough, the first imaging exam could be an ultrasonography (USG) [9, 10]. However, as around 70% of babies with congenital Zika syndrome have closed fontanel, computed tomography (CT) and magnetic resonance imaging (MRI) are preferred. CT has high sensibility to detect calcifications, is faster to be performed, and, when necessary, has shorter time of sedation, being also more available and less expensive than MRI. To its disadvantage, it uses ionizing radiation.

In its turn, MRI is the best imaging modality to evaluate brain and spinal cord malformations, very often associated findings in congenital Zika syndrome, and has demonstrated very well the brain calcifications [1]. Therefore, if it is available, it can be the first imaging modality. However, in children with an undiagnosed neurological disorder, if MRI is uninformative in regard to brain calcifications, CT should always be considered [9]. In addition, the brain damage that follows an infection by Zika virus does not necessarily lead to microcephaly, and these normocephalic infants could still present similar, but less severe, lesions when compared to the ones found in children with microcephaly [1, 11], making CT and MRI both important for investigation.

The estimated risk of developing microcephaly following maternal infection in the first trimester of pregnancy was 0.88–13.2% [12]. As in other congenital infections with microcephaly [13], we can assume that probably in this disease, also, the earlier the infection during pregnancy, the more severe the brain lesions and microcephaly. However, the initial studies have only a small sample, and we could not statistically confirm this hypothesis at the time [1]. Another limitation of the initial studies was that the symptomatology of Zika virus infection in mothers was based exclusively on their recall of skin rash during pregnancy [1]. They may not remember the exact month they had the rash, and this undermines the analysis for this hypothesis. Future prospective studies, some already in development, may confirm this hypothesis, despite the many variables involved.

Some experimental studies on mice show that Zika virus appears primarily to act on progenitor cells, resulting in cell death and disruption of neuronal proliferation, migration, and differentiation, consequently affecting neuronal viability and preventing normal brain development [14–16]. The Zika virus attacks neural progenitor cells (CNPs) as well as mature neural cells in vitro [16]. The number of CNPs decreases as the fetus develops but is still present at the end of gestation. An experimental study conducted on mice with the Brazilian Zika virus strain demonstrated that fetal infection caused intrauterine growth retardation, including signs of microcephaly [17]. In addition, this virus infection can cause a downregulation of genes involved in cell cycle pathways, dysregulation of cell proliferation, and upregulation of genes involved in apoptotic pathways resulting in cell death [14, 16].

As radiology is an important tool in the detection of congenital Zika syndrome, it is extremely necessary to recognize the major abnormalities found in imaging studies.

Prenatal Findings

Microcephaly and other abnormalities possibly related to Zika virus can be initially identified during prenatal USG. The first USG findings of congenital Zika syndrome were described in two pregnant women whose babies had fetal microcephaly [5].

The interval of time between maternal infection and identification of fetal anomalies can be from 2 to 27 weeks [5, 11, 18, 19].

Ventriculomegaly is the most frequent finding on prenatal ultrasound described in the literature (Figs. 1a and 1b) [20–22]. Calcifications are the second most common (Fig. 1a), while microcephaly is the third (Fig. 1) [22]. Microcephaly is a late finding during pregnancy, being identified from 26 to 33

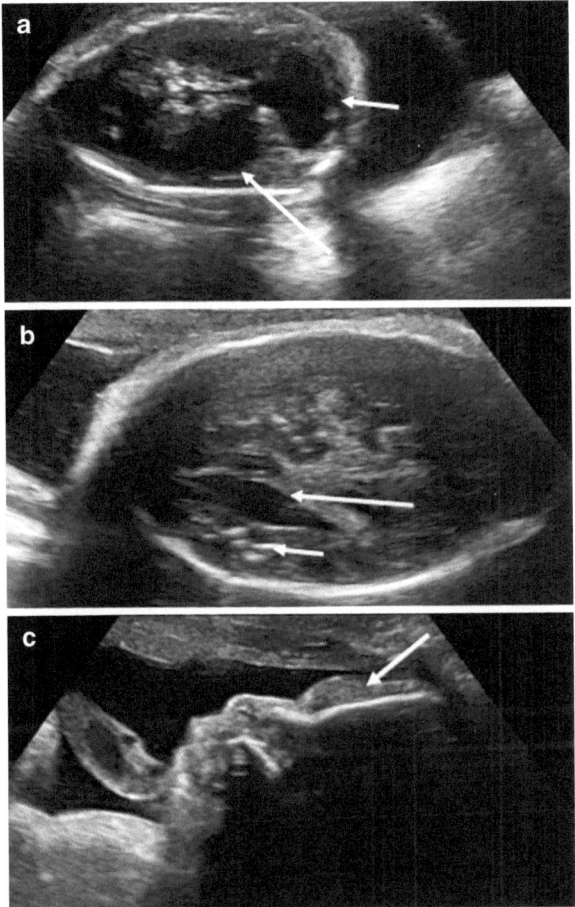

FIGURE I Prenatal USG. Axial transcerebellar image (**a**) demonstrates enlarged cisterna magna and cerebellar hypoplasia (*short arrow*) and enlarged lateral ventricle (*long arrow*). Axial image (**b**) shows ventriculomegaly (*long arrow*) and calcifications (*short arrow*). Sagittal image (**c**) shows craniofacial disproportion, inclined frontal bone, and redundant frontal skin. Courtesy of Pedro Pires, Professor of the University of Pernambuco.

gestational weeks [22]. Therefore, USG should be performed each 4 weeks since the suspected exposure, and, when this is not possible, at least once between 28 and 33 gestational weeks, because this gestational age, the fetal HC correlate better with the HC at birth [22, 23]. However, microcephaly is only confirmed after birth and can just be suspected during pregnancy [22].

Other findings are corpus callosum dysgenesis, brain reduction of volume, and enlarged cisterna magna (Fig. 1a) [5], which were identified in other studies with prenatal USG [24, 25]. The presence of increased extra-axial space and echogenic material inside the lateral ventricle that raised concern about ventricular hemorrhage is described [20]. There is also a report of septations in the lateral ventricle, usually in the occipital horns, difficult to distinguish from subventricular cysts; while subependymal cysts were occasionally visualized [24].

Postnatal Findings

Brain images of children with congenital Zika syndrome revealed that calcifications, located predominantly in the junction between the cortical and subcortical white matter, were the most frequent finding [1, 24, 26]. Malformations of cortical development are also frequently found, mostly symmetric in microcephaly cases, with simplified gyral pattern and predominance of pachygyria or polymicrogyria in the frontal lobes [1]. These two abnormalities are frequently associated; it is uncommon to find only calcifications, without malformations of cortical development (and vice-versa) in cases of congenital Zika syndrome.

Other cerebral malformations found in addition to microcephaly, which are better evaluated by MRI, are brain hypoplasia and ventriculomegaly, corpus callosum abnormalities, brainstem and cerebellum abnormalities, enlargement of the cisterna magna and of the extra-axial subarachnoid space, and delayed myelination [1]. More uncommon findings are periventricular heterotopia and cysts.

Is it possible that the cases detected so far be at the most severe end of the spectrum of brain damage caused by Zika virus? Could babies without clinical signs or symptoms from women who had the infection (clinical or subclinical) have discrete brain damage? If the responses to both are positive, these minor changes can cause future problems, such as epilepsy and cognitive and motor impairment.

The neuroimaging findings described in the literature suggest that Zika virus may disrupt various stages of the normal cortical development, with the most severe phenotypes associated with infection during the first or second trimester [27].

Volumetric MRI studies with congenital Zika syndrome have not yet been published. A preliminary analysis by the authors (not yet published) shows that the extent of the brain volume reduction seems to be better demonstrated by 3D studies than by regular MRI scans. Therefore, volumetric studies could give additional information for better characterizing the reduction of brain volume identified in these infants (Fig. 2). This information could help in understanding the physiopathology of the disease, in the differential diagnosis, and in the estimation of the prognosis for these infants.

Brain Calcifications

CT has been the mainstay of imaging for intracranial calcification for many years and remains superior for identification and delineation; however, it can be seen postnatally by transfontanellar USG (Fig. 3a and b). On CT, calcifications appear as hyperdense foci (Fig. 4). Brain calcifications in most infants predominate in the cerebral hemispheres, in the junction between the cortex and the subcortical white matter [1, 24, 26] (Fig. 5), especially in the frontal lobes [1, 26]. Other locations are the periventricular regions, basal ganglia, thalamus, and less usually the brainstem and cerebellum (Fig. 6) [1, 24, 26]. The morphology of the calcifications is varied, punctiform being the majority. Calcifications in band, coarse, or even isolated calcium spots also can be found.

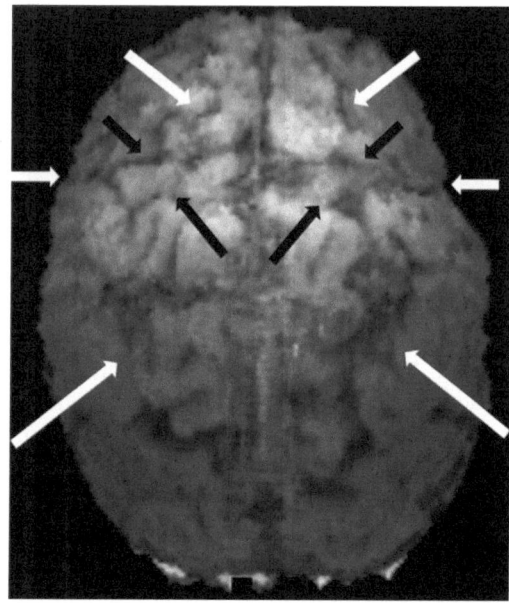

FIGURE 2 3D reconstruction of T1 spoiled gradient recalled (SPGR) shows brain reduction, mainly in the right frontal lobe, but also symmetric deep interparietal sulci, which are most evident than in conventional sectional magnetic resonance images

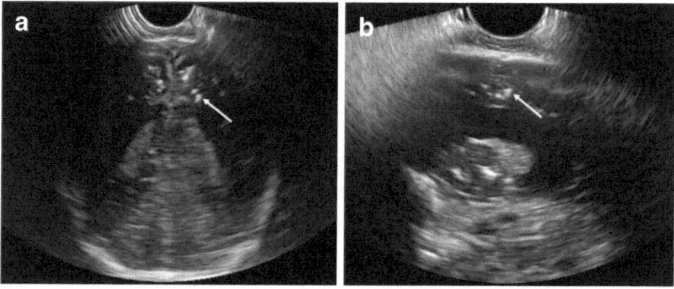

FIGURE 3 Transfontanellar US. Coronal (**a**) and sagittal (**b**) images show mild reduction of cortical thickness, ventriculomegaly, and foci of calcification in the caudate nucleus and thalamus. Courtesy of Doctor Ana Sofia Cruz

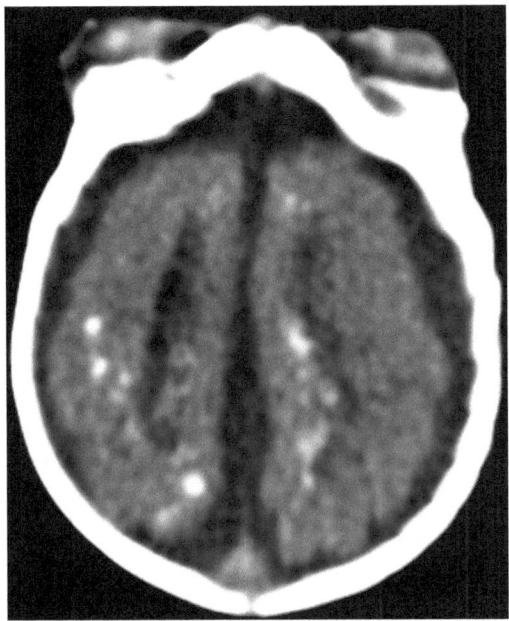

FIGURE 4 Axial CT scan of infant with confirmed congenital Zika syndrome. Multiple subcortical and periventricular hyperdense foci of calcifications

Among all MRI techniques, susceptibility magnetic weighted imaging is the best sequence to demonstrate intracranial calcifications [9]. In our experience, MRI was able to detect calcifications, and the majority were seen as hyperintense foci on T1-weighted images (Fig. 7) and hypointense on susceptibility magnetic weighted and T2* gradient echo images (Fig. 8) [1]. Differential diagnoses for hyperintensity on T1-weighted images are, for example, methemoglobin, manganese deposition, melanin, and lipid [9, 28]. Differential diagnoses for hypointensity on T2* [9, 28] and susceptibility magnetic weighted images [29] are iron deposition, iron deposition, deoxyhemoglobin, hemosiderin/ferritin, and small veins in susceptibility magnetic weighted images [29].

The exact mechanism of how or why this causal factor (congenital Zika virus infection) is now linked to such severe

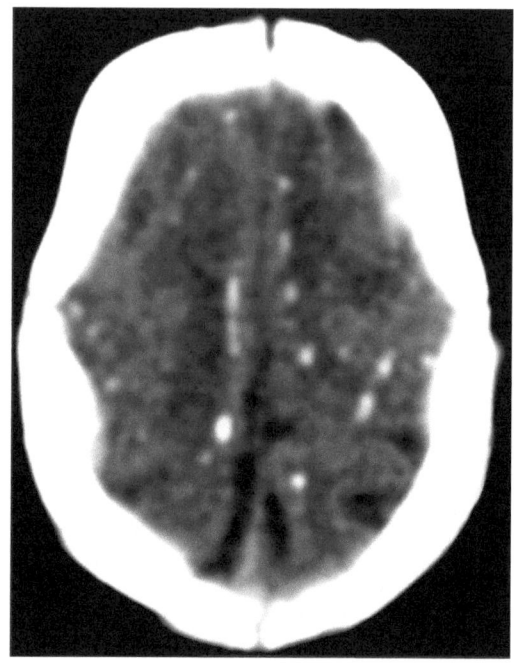

FIGURE 5 Axial CT scan of infant with confirmed congenital Zika syndrome. Multiple hyperdense foci in the junction between the cortex and the subcortical white matter

brain damage in infants is yet unknown. The hypothesis is that virus strains in Brazil [17] lead to the destruction of brain cells, forming many lesions similar to "scars" in which calcium is deposited over time. The location of the calcifications at the gray matter-white matter interface could suggest a vascular component to the infection [24].

Surprisingly, follow-up evaluations, around one year after birth, have been showing reduction in size and number of calcifications and further reduction in parenchymal thickness (see Chap. 8, Fig. 17, and Chap. 9, Fig. 11). The reason why this is happening is yet to be discovered; however, we suppose that it could represent evolution of the disease, for example, by direct virus damage or immunologic reaction.

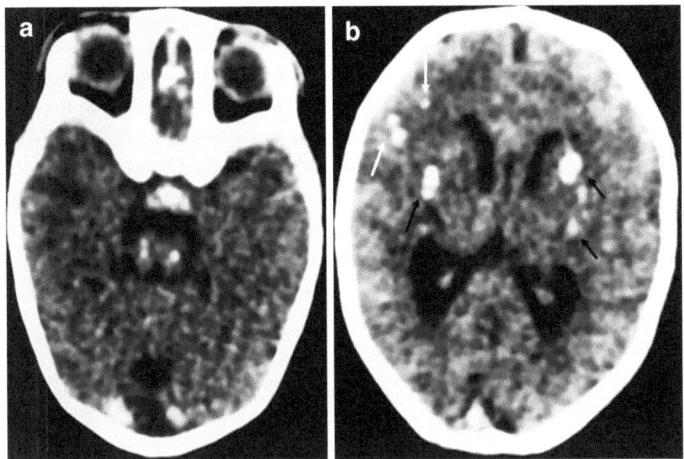

FIGURE 6 Axial CT scan of infant with probable congenital Zika syndrome. Bilateral hyperdense foci in the midbrain (a), coarse calcification in the basal ganglia (*black arrows* in **b**), and punctiform calcification in the junction between the cortex and the subcortical white matter in the right frontal lobe (*white arrow* in **b**)

Malformations of Cortical Development

Malformations of cortical development (MCDs) present in congenital Zika syndrome are believed to be related to the death of cortical progenitor cells caused by the virus [16]. The major MCDs, better evaluated by MRI scans, are simplified gyral pattern, pachygyria, and polymicrogyria [1]. In congenital Zika syndrome, polymicrogyria and pachygyria are predominant in the frontal lobes, while simplified gyral pattern is found more parietal and occipital lobes or diffusely [1].

Simplified gyral pattern is related to abnormal neuronal and glial proliferation, with deficits in cell production or white matter development, while pachygyria is related to abnormal neuronal migration [30]. Simplified gyral pattern is defined as few gyri and shallow sulci (decreased grooves

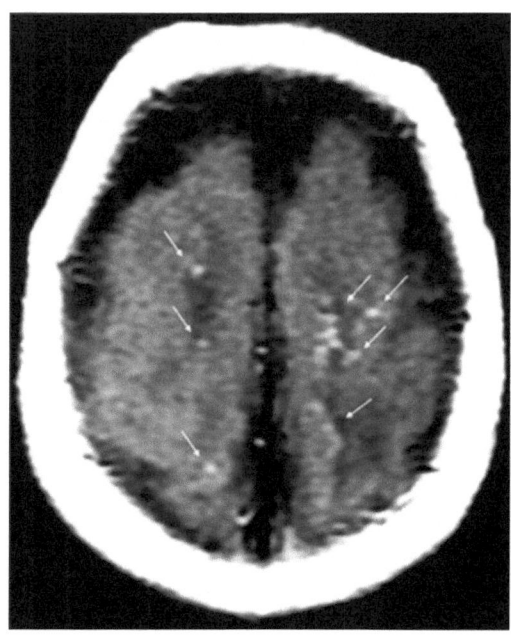

FIGURE 7 Brain MRI of infant with confirmed congenital Zika syndrome. Axial T1-weighted image shows multiple foci of high signal intensity in the junction between cortical and subcortical white matter disease (*white arrows*)

and spins), usually identified in children with microcephaly, but the cortex is not thickened in these areas (Fig. 9) [31].

Meanwhile, pachygyria refers to areas of increased cortical thickness and broad, flat gyri, also known as incomplete lissencephaly (Fig. 10) [32]. It is a malformation that originates early in pregnancy, between the 12th and 16th gestational week, and results from arrested neuronal migration [32]. The lissencephaly is considered complete when there is agyria ("smooth brain") and thickened cortex (microlissencephaly) (Fig. 11) [32]. The presence of pachygyria suggests early fetal infection [13].

Since simplified gyral pattern can also be mistaken for broad gyri, the feature that differentiates it from pachygyria

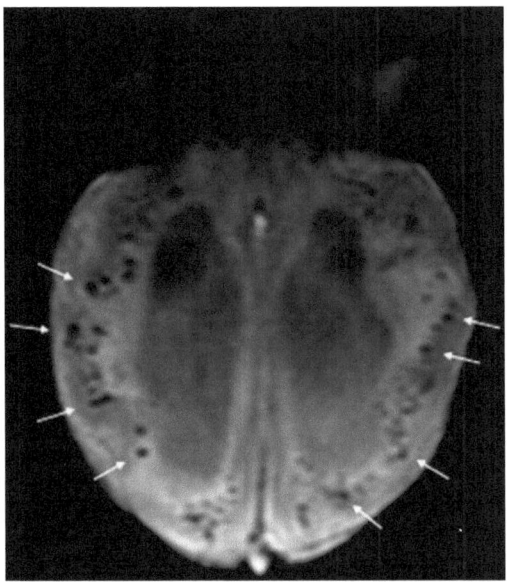

FIGURE 8 Brain MRI of infant with confirmed congenital Zika syndrome. Axial SWI shows multiple foci of low signal intensity in the junction between cortical and subcortical white matter disease (*white arrows*) and ventriculomegaly

is cortical thickness (normal or thin cortex in simplified gyral pattern and thick cortex in pachygyria) [31]. When both pachygyria and simplified gyral pattern are present in one infant with Zika virus congenital infection, the former is always located anteriorly to the latter (e.g., pachygyria in the frontal lobes and simplified gyral pattern in the parietal and occipital lobes).

Polymicrogyria is a malformation related to an interruption in late stages of neuronal migration and cortical organization [30], which originates only after the 20th gestational week [32]. It is identified through thickened cortex and irregular cortical surface and cortico-subcortical junction due to the presence of numerous microgyria and microsulci (Fig. 12) [33].

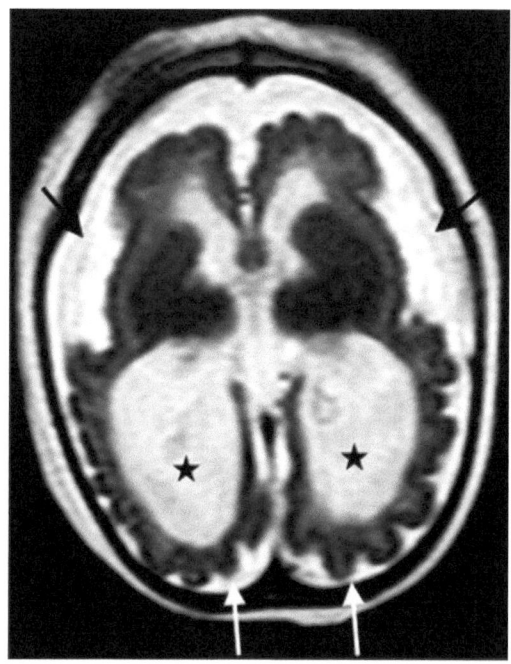

FIGURE 9 Brain MRI of infant with confirmed congenital Zika syndrome infection. Axial T2-weighted image shows simplified gyral pattern (*white arrows*) and widely open Sylvius fissures (*black arrows*), enlarged extra-axial subarachnoid frontal and temporal space, and ventriculomegaly (*stars*). Reproduced with permission from de Fátima Vasco Aragão et al. [1]

In our experience, pachygyria-agyria complex and simplified gyral pattern are more severe malformations, leading to reduction of brain parenchyma volume and, consequently, ventriculomegaly and enlarged extra-axial CSF space. These malformations have been only found in infants with microcephaly, with or without arthrogryposis [34]. In infants with arthrogryposis, pachygyria and simplified gyral pattern were the only malformation found; polymicrogyria is absent in all of these cases [34].

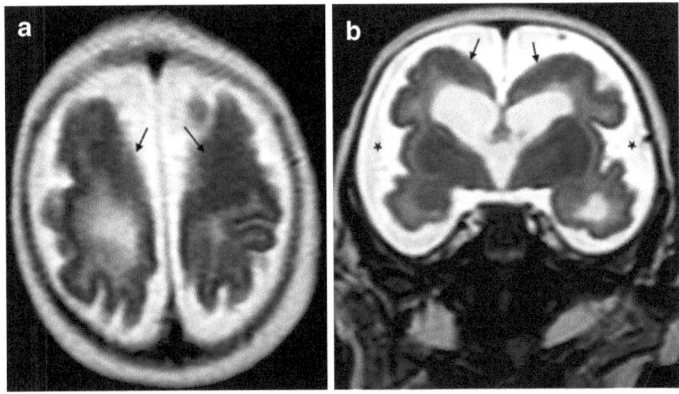

FIGURE 10 Brain MRI of infant with confirmed congenital Zika syndrome. Axial T2-weighted image (**a**) shows microcephaly and pachygyria in the frontal lobes (*black arrows*). Note the cortical thickness in the pachygyric frontal lobe (*black arrows*), shown on coronal T2-weighted images (**b**) and bilateral diffuse enlargement of subarachnoid space. Reproduced with permission from de Fátima Vasco Aragão et al. [1]

In infants born without microcephaly, polymicrogyria was the only malformation found; pachygyria and simplified gyral pattern were absent in all of these cases (not yet published). Polymicrogyria is probably an anatomically less severe lesion, preserving the parenchyma thickness, with consequent preserved brain volume and corpus callosum (not yet published). Although functional outcome cannot be predicted, it is possible that the neuropsychomotor impairment will be less severe in infants with this malformation (not yet published).

Brain Hypoplasia and Ventriculomegaly

As malformations of cortical development, brain hypoplasia is probably a consequence of the arrest in development caused by Zika virus. It can be perceived as decreased brain volume, reduced white matter thickness, and enlarged

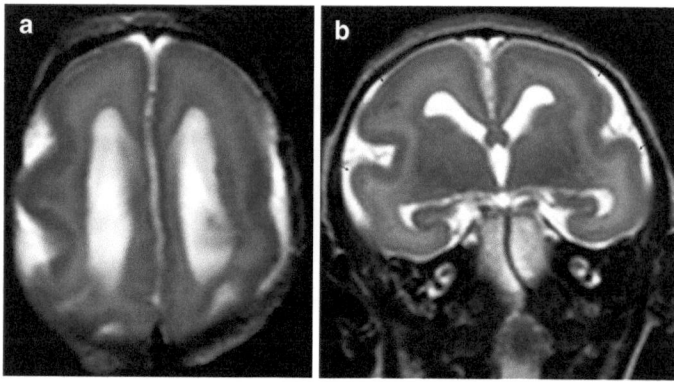

FIGURE 11 Brain MRI of infant with microlissencephaly and probable congenital Zika syndrome. Axial T2-weighted image (**a**) shows almost completely smooth cerebral surface with a diffuse thick cortex on coronal T2-weighted images (**b**). Reproduced with permission from de Fátima Vasco Aragão et al. [1]

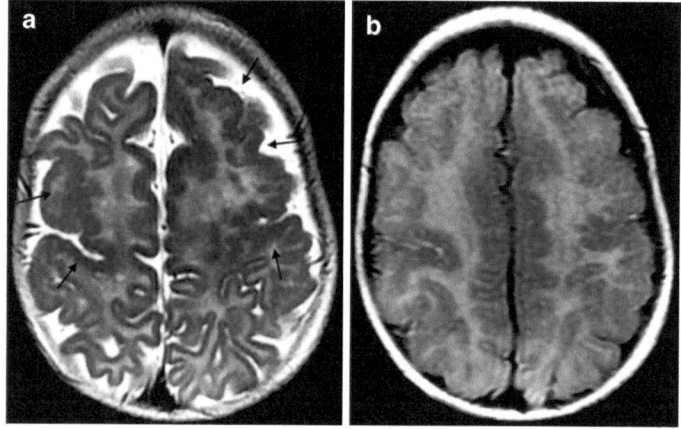

FIGURE 12 Brain MRI of infant with probable congenital Zika syndrome. Axial T2-weighted image (**a**) shows bilateral frontal and central sulcus polymicrogyria (*black arrows*). Note the thickened and irregular cortical-white matter junction on FLAIR-weighted image (**b**). Reproduced with permission from de Fátima Vasco Aragão et al. [1]

ventricles (ventriculomegaly) [1]. Ventriculomegaly, when present, is characterized by predominant enlargement of the posterior portions of the lateral ventricles (trigones and the posterior horns), associated with the hypoplastic or hypogenetic corpus callosum, but usually involves the whole ventricular system (Fig. 13) [1, 24].

Severe ventriculomegaly, due to hydrocephalus, and even a prominent extra-axial subarachnoid space can make the head circumference appear bigger than the actual volume, thereby masking a small microcephalic brain [1]. Some infants have obstruction of the cerebrospinal fluid pathways and/or reduced absorption of cerebrospinal fluid, leading to hydrocephalus.

Corpus Callosum Abnormalities

Malformations of the corpus callosum may also be present, being a frequent finding in moderate to severe cases of microcephaly [1]. It can be either hypogenetic (Fig. 14), when it is not completely formed, or hypoplastic (Fig. 15), when all its portions are present, but its thickness is decreased [31].

Similar to what has been described for human immunodeficiency virus and herpesvirus, the corpus callosum changes may be associated with a decreased number of neuronal cells and/or the interference of the virus in the neuronal migration process [11, 35, 36].

Brainstem and Cerebellum Abnormalities

Although not as frequent as brain abnormalities, calcifications and hypoplasia can be found in the brainstem, mainly at the pons (Fig. 16), and in the cerebellum (Fig. 17) of infants with congenital Zika syndrome [1]. One possible explanation is that it can be due to Wallerian degeneration and/or to arrested development of the corticospinal and corticobulbar tracts, as well as pontocerebellar connections [37].

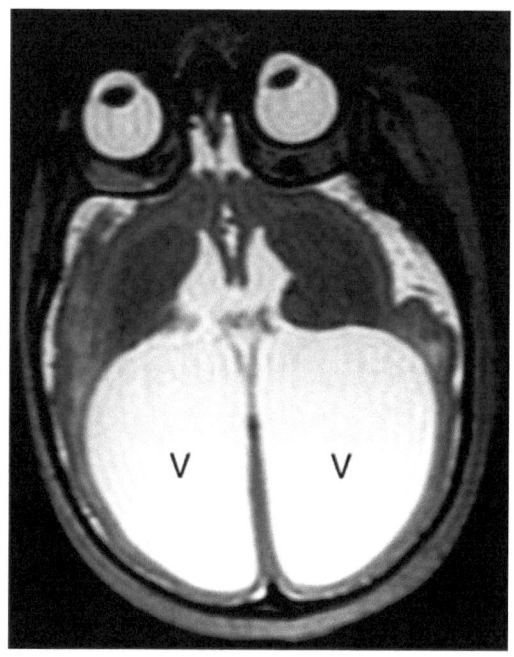

FIGURE 13 Brain MRI of infant with confirmed congenital Zika syndrome. Axial T2-weighted image shows extremely simplified gyral pattern with thin cortex and severe ventriculomegaly, mainly at the posterior horn and ventricular atrium (*V*). Note the bulging walls of the ventricles

Melo and colleagues [11] observed a slender brainstem on fetal MRI in very severely affected infants that died within hours after born, making it impossible for the fetus to breathe normally at birth.

Enlarged Cisterna Magna and Extra-Axial Subarachnoid Space

Cisterna magna and extra-axial subarachnoid space can both be enlarged in cases of congenital Zika syndrome. Enlarged cisterna magna is defined as a cisterna magna measuring

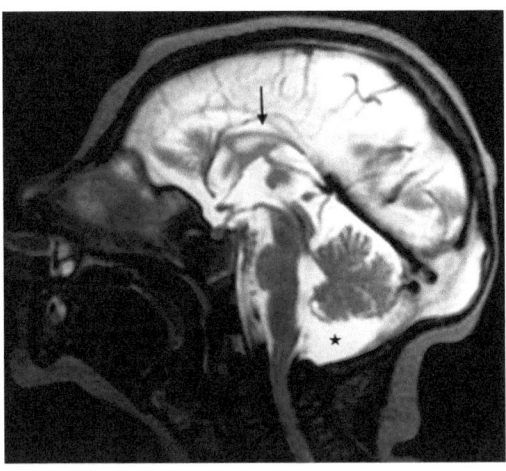

FIGURE 14 Brain MRI of infant with confirmed congenital Zika syndrome. Sagittal T2-weighted image shows hypogenetic corpus callosum (*black arrow*) and enlarged cistern magna (*star*)

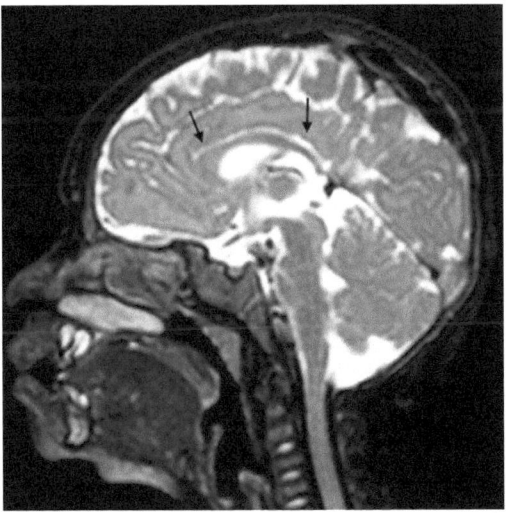

FIGURE 15 Brain MRI of infant with confirmed congenital Zika syndrome. Sagittal T2-weighted image shows hypoplastic corpus callosum (*black arrows*) and enlarged cisterna magna

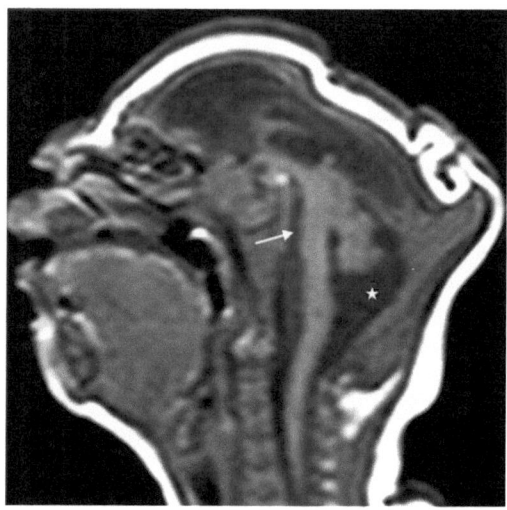

FIGURE 16 Brain MRI of infant with severe microcephaly and confirmed congenital Zika syndrome. Sagittal T1-weighted image shows severe pons hypoplasia (*white arrow*). Note enlarged cisterna magna (*star*) and cerebellum hypoplasia. Reproduced with permission from de Fátima Vasco Aragão et al. [1]

≥10 mm in midsagittal planes (Fig. 18) [38], while enlarged extra-axial subarachnoid space is determined qualitatively by visual inspection (Fig. 19).

Although these findings can be considered normal variations, their frequency in infants with other Zika-related lesions, especially in the most severe cases, is very high [1]. They can be indirect signs of the overall hypoplasia found in these infants; however, some cases present enlarged cisterna magna and enlarged extra-axial subarachnoid space even without cerebellum and brain hypoplasia, respectively.

Delayed Myelination

Some infants with congenital Zika syndrome have impairments in myelination that can be identified on MRI. The myelination process can be perceived as changes in the signal

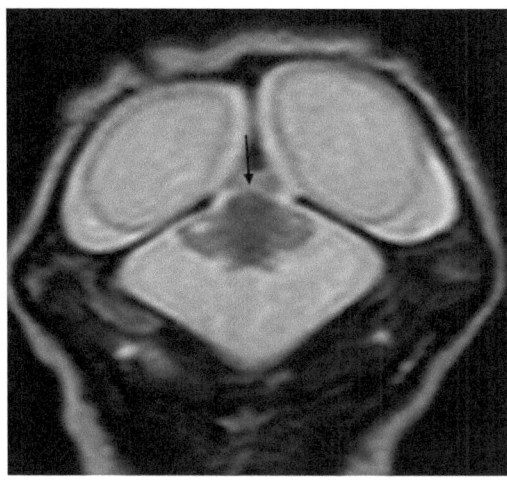

FIGURE 17 Brain MRI of infant with confirmed congenital Zika syndrome. Coronal T2-weighted image shows severe cerebellar hypoplasia (*black arrow*). Additionally, note an enlargement of extra-axial subarachnoid space supra- and infratentorial and ventriculomegaly. Reproduced with permission from de Fátima Vasco Aragão et al. [1]

intensity, which, in general, progress from the center of the brain to the periphery, from inferior to superior, and from dorsal to ventral [31].

Some milestones are used to evaluate the progression of myelination, which can be identified as changes for high and low signal intensity, respectively, on T1- and T2-weighted images [31]. For example, on T1, the changes are seen in the anterior limbs of internal capsule by 3 months of age, in the splenium of the corpus callosum by 4 months, in the genu of the corpus callosum by 6 months, and, at 8 months, the brain presents the adult pattern, except in the most subcortical fibers [31]. On T2, the changes in signal intensity are seen in the splenium of the corpus callosum by 6 months, in the genu of the corpus callosum by 8 months, in the anterior limbs of internal capsule by 11 months, and, at 18 months, the brain presents the adult pattern, except for the subcortical white matter [31].

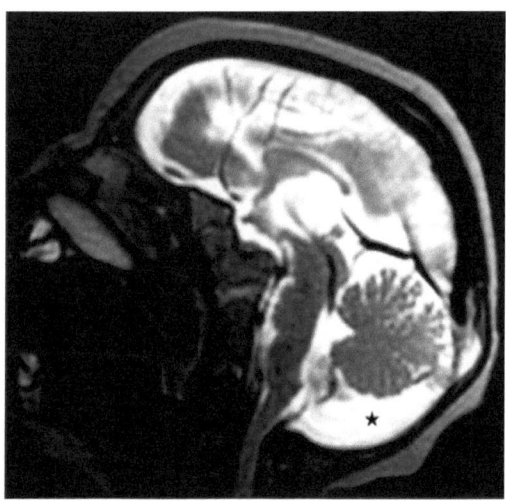

FIGURE 18 Brain MRI of infant with confirmed congenital Zika syndrome. Sagittal T2-weighted image shows enlarged cistern magna (*star*). Note the hypogenetic corpus callosum and pons

Vascular Abnormalities

Heterogeneous material in the confluence of the sinuses, detected by US and CT, was reported in infants with congenital Zika syndrome by Soares de Oliveira-Szejnfeld and colleagues [24]. This finding was believed to correspond to thrombus [24]. In our experience, it is difficult to evaluate the frequency of this finding, as contrasted studies are not routinely performed on these infants. However, in the only infant who underwent contrasted MRI in our data, impregnation of the dura mater and a filling defect of the superior sagittal sinus were found, which could indicate thrombosis (Fig. 20).

Reduced Thickness of Spinal Cord and Ventral Roots

Arthrogryposis is the most severe consequence of the Zika virus damage to the spinal cord. It is a syndrome characterized by joint contractures, present since birth, affecting two

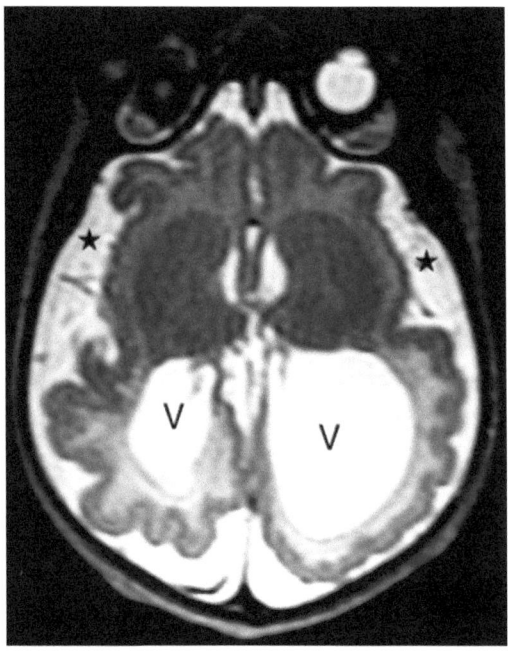

FIGURE 19 Brain MRI of infant with confirmed congenital Zika syndrome. Axial T2-weighted image shows enlarged extra-axial subarachnoid space (*stars*), moderate asymmetrical ventriculomegaly (*V*) and malformation of cortical development

or more areas of the body [39, 40]. Children affected by arthrogryposis had onset and severe weakness early during intrauterine life, with immobilization of joints at different developmental stages [40]. Arthrogryposis multiplex congenita can be caused by many different disorders as defects of the uterine environment, disorders of the connective tissues, muscular dystrophies, and other neurogenic abnormalities or conditions that affect the central or peripheral nervous systems in at least one of the components of the motor pathways, from the spinal cord to muscles [40]. However, in congenital Zika syndrome, we have suggested a neurogenic cause for arthrogryposis, because the major imaging abnormalities found were reduced thickness of the spinal cord (Fig. 21) and reduction of the anterior

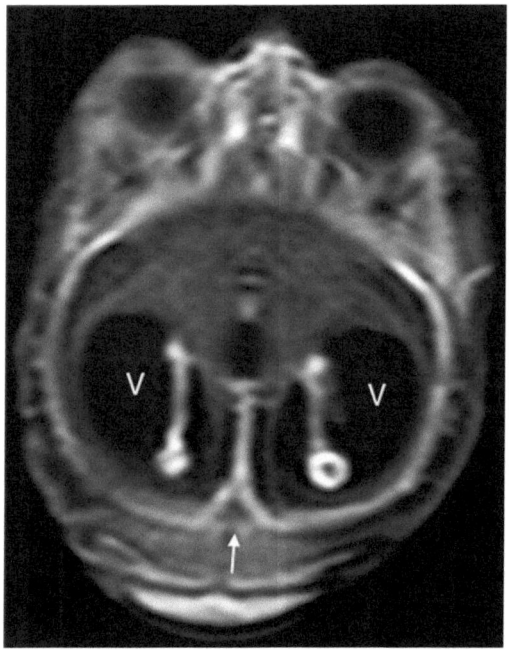

FIGURE 20 Brain MRI of infant with severe microcephaly and confirmed congenital Zika syndrome. Axial T1-weighted image fat suppression post-contrast shows filling defect within superior sagittal sinus (*white arrow*) and thickness and enhancement of frontal dura mater and severe ventriculomegaly (*V*). Reproduced with permission from de Fátima Vasco Aragão et al. [1]

nerve roots of the medullary cone (Fig. 22) [4]. These findings are not restricted to infants with arthrogryposis, but there is a spectrum of abnormalities in the spinal cord and ventral roots in Zika virus congenital syndrome even in patients without arthrogryposis [34].

In fact, most of the babies with congenital Zika virus syndrome analyzed had some degree of spinal cord reduction of thickness [34]. This reduction is predominant in the thoracic segment in cases without arthrogryposis and in the whole spinal cord in cases with arthrogryposis [34].

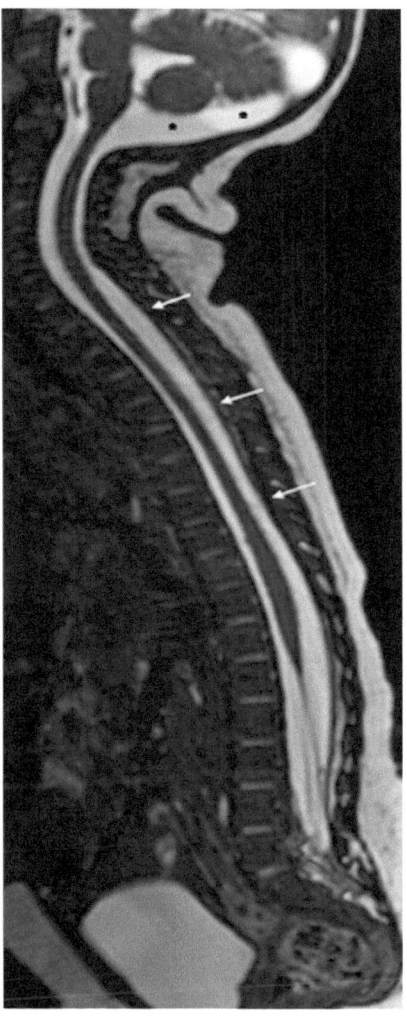

FIGURE 21 Spine MRI of infant with confirmed congenital Zika syndrome and arthrogryposis. Sagittal T2-weighted fast imaging employing steady-state acquisition (FIESTA) showing apparently reduced spinal cord thickness (*short arrows*). Mega cisterna magna is seen (*stars*). Reproduced with permission from Van der Linden et al. [4]

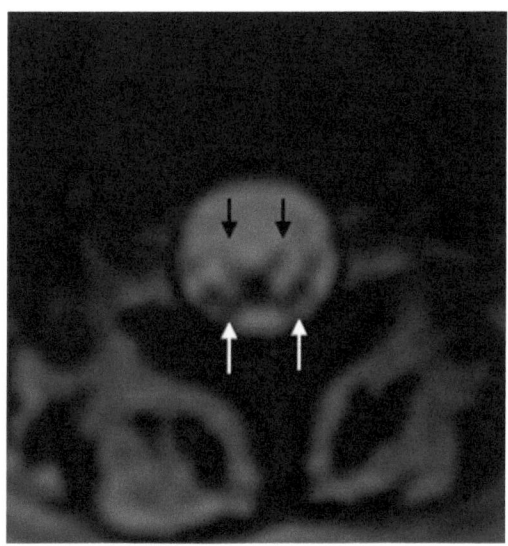

FIGURE 22 Spine MRI of infant with confirmed congenital Zika syndrome and arthrogryposis. Axial reconstruction of T2-weighted FIESTA showing reduction of medullary cone ventral roots (*black arrows*) compared with dorsal roots (*white arrows*). Reproduced with permission from de Fátima Vasco Aragão et al. [34]

The reduction of anterior nerve roots of the conus medullaris is also present in infants without arthrogryposis, although it is less severe [34]. In the brain, periventricular calcifications were shown to be more frequent in patients with arthrogryposis [34].

References

1. de Fátima Vasco Aragão M, van der Linden V, Brainer-Lima AM, Coeli RR, Rocha MA, Silva PS d, Carvalho MDCG d, Linden A v d, Holanda AC d, Valenca MM. Clinical features and neuroimaging (CT and MRI) findings in presumed Zika virus

related congenital infection and microcephaly: retrospective case series study. BMJ. 2016;353:i1901.

2. Leal M de C, Muniz LF, Caldas Neto S da S, van der Linden V, Ramos RCF. Sensorineural hearing loss in a case of congenital Zika virus. Braz J Otorhinolaryngol. 2016. doi:10.1016/j.bjorl.2016.06.001.

3. Ventura CV, Maia M, Ventura BV, et al. Ophthalmological findings in infants with microcephaly and presumable intra-uterus Zika virus infection. Arq Bras Oftalmol. 2016;79:1–3.

4. Linden V v d, Filho ELR, Lins OG, et al. Congenital Zika syndrome with arthrogryposis: retrospective case series study. BMJ. 2016;354:i3899.

5. Oliveira Melo AS, Malinger G, Ximenes R, Szejnfeld PO, Alves Sampaio S, Bispo de Filippis AM. Zika virus intrauterine infection causes fetal brain abnormality and microcephaly: tip of the iceberg? Ultrasound Obstet Gynecol. 2016;47:6–7.

6. França GVA, Schuler-Faccini L, Oliveira WK, et al. Congenital Zika virus syndrome in Brazil: a case series of the first 1501 live-births with complete investigation. Lancet Lond Engl. 2016;388:891–7.

7. Ventura CV, Maia M, Dias N, Ventura LO, Belfort R. Zika: neurological and ocular findings in infant without microcephaly. Lancet Lond Engl. 2016;387:2502.

8. Ministério da Saúde (Brazil). Protocolo de vigilância e resposta à ocorrência de microcefalia e/ou alterações do sistema nervoso central (SNC). 2016 http://portalsaude.saude.gov.br/images/pdf/2016/janeiro/22/microcefalia-protocolo-devigilancia-e-resposta-v1–3-22jan2016.pdf. Accessed 16 May 2016.

9. Livingston JH, Stivaros S, Warren D, Crow YJ. Intracranial calcification in childhood: a review of aetiologies and recognizable phenotypes. Dev Med Child Neurol. 2014;56:612–26.

10. World Health Organization. Avaliação de bebês com microcefalia no contexto do vírus Zika. 2016. http://apps.who.int/iris/bitstream/10665/204475/8/WHO_ZIKV_MOC_16.3_por.pdf. Accessed 28 Dec 2016.

11. Melo AS d O, Aguiar RS, MMR A, et al. Congenital Zika virus infection: beyond neonatal microcephaly. JAMA Neurol. 2016;73:1407–16.

12. Johansson MA, Mier-y-Teran-Romero L, Reefhuis J, Gilboa SM, Hills SL. Zika and the risk of microcephaly. N Engl J Med. 2016;375:1–4.

13. Fink KR, Thapa MM, Ishak GE, Pruthi S. Neuroimaging of pediatric central nervous system cytomegalovirus infection. Radiographics. 2010;30:1779–96.
14. Garcez PP, Loiola EC, Madeiro da Costa R, Higa LM, Trindade P, Delvecchio R, Nascimento JM, Brindeiro R, Tanuri A, Rehen SK. Zika virus impairs growth in human neurospheres and brain organoids. Science. 2016;352:816–8.
15. Li C, Xu D, Ye Q, Hong S, Jiang Y, Liu X, Zhang N, Shi L, Qin C-F, Xu Z. Zika virus disrupts neural progenitor development and leads to microcephaly in mice. Cell Stem Cell. 2016;19:672.
16. Tang H, Hammack C, Ogden SC, et al. Zika virus infects human cortical neural progenitors and attenuates their growth. Cell Stem Cell. 2016;18:587–90.
17. Cugola FR, Fernandes IR, Russo FB, et al. The Brazilian Zika virus strain causes birth defects in experimental models. Nature. 2016;534:267–71.
18. Mlakar J, Korva M, Tul N, et al. Zika virus associated with microcephaly. N Engl J Med. 2016;374:951–8.
19. Calvet G, Aguiar RS, Melo ASO, et al. Detection and sequencing of Zika virus from amniotic fluid of fetuses with microcephaly in Brazil: a case study. Lancet Infect Dis. 2016;16:653–60.
20. Driggers RW, Ho C-Y, Korhonen EM, et al. Zika virus infection with prolonged maternal viremia and fetal brain abnormalities. N Engl J Med. 2016;374:2142–51.
21. Zare Mehrjardi M, Keshavarz E, Poretti A, Hazin AN. Neuroimaging findings of Zika virus infection: a review article. Jpn J Radiol. 2016;34:765–70.
22. Vouga M, Baud D. Imaging of congenital Zika virus infection: the route to identification of prognostic factors. Prenat Diagn. 2016;36:799–811.
23. Baud D, Van Mieghem T, Musso D, Truttmann AC, Panchaud A, Vouga M. Clinical management of pregnant women exposed to Zika virus. Lancet Infect Dis. 2016;16:523.
24. Soares de Oliveira-Szejnfeld P, Levine D, Melo AS d O, et al. Congenital brain abnormalities and Zika virus: what the radiologist can expect to see prenatally and postnatally. Radiology. 2016;281:203–18.
25. Carvalho FHC, Cordeiro KM, Peixoto AB, Tonni G, Moron AF, Feitosa FEL, Feitosa HN, Araujo Júnior E. Associated ultrasonographic findings in fetuses with microcephaly because of suspected Zika virus (ZIKV) infection during pregnancy. Prenat Diagn. 2016;36:882–7.

26. Hazin AN, Poretti A, Cruz DDCS, et al. Computed tomographic findings in microcephaly associated with Zika virus. N Engl J Med. 2016;374:2193–5.

27. Brasil P, Pereira Jr JP, Raja Gabaglia C, et al. Zika virus infection in pregnant women in Rio de Janeiro—preliminary report. N Engl J Med. 2016;0:null.

28. Hess C, Purcell D. Analysis of density, signal intensity and echogenicity. In: Naidich et al. Imaging of the Brain. 1st ed. Philadelphia: Saunders; 2013:45–66.

29. Haacke EM, Mittal S, Wu Z, Neelavalli J, Cheng Y-CN. Susceptibility-weighted imaging: technical aspects and clinical applications, Part 1. Am J Neuroradiol. 2009;30:19–30.

30. Barkovich AJ, Guerrini R, Kuzniecky RI, Jackson GD, Dobyns WB. A developmental and genetic classification for malformations of cortical development: update 2012. Brain. 2012;135(Pt 5):1348–69.

31. Barkovich AJ, Raybaud C. Pediatric neuroimaging. Philadelphia, PA: Lippincott Williams & Wilkins; 2012.

32. Barkovich AJ, Gressens P, Evrard P. Formation, maturation, and disorders of brain neocortex. Am J Neuroradiol. 1992;13:423–46.

33. Barkovich AJ, Hevner R, Guerrini R. Syndromes of bilateral symmetrical polymicrogyria. Am J Neuroradiol. 1999;20:1814–21.

34. de Fátima Vasco Aragão M, Brainer-Lima AM, Holanda AC, van der Linden V, Aragão LV, Silva Júnior MLM, Sarteschi C, Petribu NCL, van der Linden A, Valença MM. (IN PRESS) Spectrum of spinal cord, spinal roots and brain MRI abnormalities in congenital Zika syndrome with and without arthrogryposis. AJNR. 2017.

35. Corey L, Wald A. Maternal and neonatal herpes simplex virus infections. N Engl J Med. 2009;361:1376–85.

36. Bosnjak VM, Daković I, Duranović V, Lujić L, Krakar G, Marn B. Malformations of cortical development in children with congenital cytomegalovirus infection—a study of nine children with proven congenital cytomegalovirus infection. Coll Antropol. 2011;35(Suppl 1):229–34.

37. Štrafela P, Vizjak A, Mraz J, Mlakar J, Pižem J, Tul N, Županc TA, Popović M. Zika virus-associated micrencephaly: a thorough description of neuropathologic findings in the fetal central nervous system. Arch Pathol Lab Med. 2016. doi:10.5858/arpa.2016-0341-SA.

38. Bosemani T, Orman G, Boltshauser E, Tekes A, Huisman TAGM, Poretti A. Congenital Abnormalities of the Posterior Fossa. Radiographics. 2015;35:200–20.
39. Bamshad M, Van Heest AE, Pleasure D. Arthrogryposis: a review and update. J Bone Joint Surg Am. 2009;91:40–6.
40. Banker BQ. Arthrogryposis multiplex congenita: spectrum of pathologic changes. Hum Pathol. 1986;17:656–72.

Radiological Differential Diagnosis of Microcephaly

Natacha Calheiros de Lima Petribu
and Maria de Fátima Viana Vasco Aragão

The cranioencephalic abnormalities caused by Zika virus congenital infection, although following a pattern, are not pathognomonic and are common to different etiologies.

The imaging spectrum varies from normal to quite altered, with the following main findings: calcifications in the cortico-subcortical junction, basal ganglia, and thalamus; cerebral and cerebellar hypoplasia; ventriculomegaly (mainly colpocephaly); corpus callosum agenesis; neuronal migration defects; and prominent occipital bone [1–4].

Congenital microcephaly is also called primary microcephaly, since it is present at birth, and postnatal is called secondary microcephaly. Both primary and secondary microcephaly may be genetic or acquired in origin. The distinction between primary and secondary microcephaly is important for establishing an appropriate differential diagnosis. We will approach only primary microcephaly, since it

N.C. de Lima Petribu (✉)
Hospital Barão de Lucena, Recife, PE, Brazil
e-mail: natachacalheiros@yahoo.com.br

M.F.V.V. Aragão
Centro Diagnóstico Multimagem, Mauricio de Nassau University, Recife, PE, Brazil
e-mail: fatima.vascoaragao@gmail.com

M.F.V.V. Aragão (ed.), *Zika in Focus*,
DOI 10.1007/978-3-319-53643-9_7,
© Springer International Publishing AG 2017

represents most of the described cases of congenital infection by Zika virus. There are ongoing research and reports of neonates who were born normocephalic and who progressed with a fall in the percentile of the head circumference (HC) and later became microcephalic. The pathophysiology in these cases is still unclear, and the main hypotheses are latent intrauterine infection that manifested in late gestation or infection occurred at the end of gestation with few changes at the time of birth or if there was postnatal infection.

Von der Hagen (2014) proposed a microcephaly investigation flowchart that includes family history, physical examination, specific tests (metabolic and genetic), transfontanellar ultrasound, and magnetic resonance imaging (MRI) [5]. We believe that computed tomography (CT) without contrast has a very important role in the diagnostic investigation, since it is the best method to evaluate intracranial calcifications and the gold standard to evaluate sutures and fontanelles (3D reconstruction).

We must suspect primary microcephaly of genetic origin when there is a family history (blood relatives, other affected family members) or when there is a phenotype characteristic of some syndrome. Among the genetic causes of primary microcephaly are trisomies 13, 18, and 21, Cornelia de Lange syndrome, Rett syndrome, Angelman syndrome, and metabolic diseases, among others [5]. This list is quite extensive, and these pathologies, in general, are less frequent than those acquired.

Aicardi-Goutières syndrome (AGS) and RNASET2-related disease show brain calcifications and cysts in the temporal lobe, but cortical malformations have not been described in these conditions. AGS is a type 1 interferonopathy that is caused by mutations in any one of the genes TREX1, RNASEH2A, RNASEH2B, RNASEH2C, SAMHD1, and ADAR1. There are patients with AGS who do not have mutations in any of these genes. The typical AGS neonatal presentation is frequently characterized by fever, seizures, hepatosplenomegaly, thrombocytopenia, and anemia, with or without microcephaly [6] (Fig. 1).

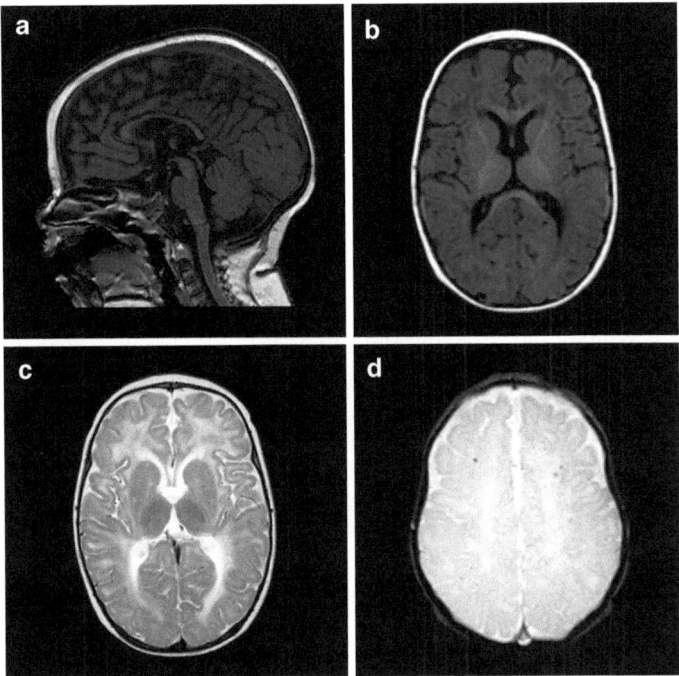

FIGURE I MR images of a term 13-month-old boy with TREX1 gene mutation and Aicardi-Goutières syndrome presented with global developmental delay; no words; unable to sit independently, roll, or walk; and spasticity. (**a**) Sagittal T1: thin corpus callosum for age. (**b**) Axial T1: some myelination in internal capsule, otherwise abnormal low signal on T1 s throughout white matter consistent with hypomyelination. (**c**) Axial T2: scant myelination in posterior limbs of internal capsule, otherwise lack of expected low signal on T2 throughout the white matter. (**d**) Susceptibility: low signal foci of calcification in cortico-subcortical white matter junction mainly in frontal lobes. Image courtesy of Doctors Paul Caruso and Ganeshwaran Mochida (Massachusetts General Hospital, Boston, USA)

Another genetic disease, band-like calcification (BLC) with simplified gyration and polymicrogyria, presents a characteristic pattern of symmetrical continuous or semicontinuous ribbon of cortical calcification. BLC is caused by mutations in the OCLN gene, encoding the tight junction protein occludin.

BLC is a severe disorder presenting at birth with seizures, feeding difficulties, quadriplegia, and bulbar palsy. There is a generalized malformation of the cerebrum with very primitive sulcation and abnormal gyration with a frontoparietal predominance. Most patients present hypoplasia of the cerebellum, brain stem, and corpus callosum. There is a marked reduction in white matter volume, with abnormal high T2-weighted signal and a lack of normal myelination. There is also symmetrical thalamic and central pontine calcification [6].

Among the etiologies of primary acquired microcephaly are teratogenic agents (alcohol, cocaine, anticonvulsant drugs, mercury poisoning, and radiation), vascular incident, maternal disease (hyperphenylalaninemia and anorexia), placental insufficiency, and craniosynostosis. The main cause of primary microcephaly of acquired origin is intrauterine infections. Many infectious agents can compromise the fetal intrauterine central nervous system (CNS), with the most common agents being STORCH (syphilis, toxoplasmosis, rubella, cytomegalovirus, and herpes virus) [7]. We will describe the most frequent ones.

Cytomegalovirus (CMV)

CMV infection is the most common congenital viral infection and can affect up to 1% of live births. Prenatal infection occurs through transplacental transmission. Many newborns with congenital infection appear normal, but approximately 10% will be symptomatic at birth. Infection in symptomatic infants ranges from mild to severe life-threatening disseminated disease, accounting for up to 20% of perinatal mortality. More than 80% of symptomatic newborns may present sequels such as intellectual disability, cerebral palsy, convulsions, visual problems, and sensorineural hearing loss [7, 8].

Approximately 68% of symptomatic neonates present normal unenhanced head CT. Microcephaly is the most specific predictive factor of mental retardation and motor disturbance, while changes in head CT are the most sensitive predictive factor. There is a correlation between HC and intelligence coefficient (IQ) [9].

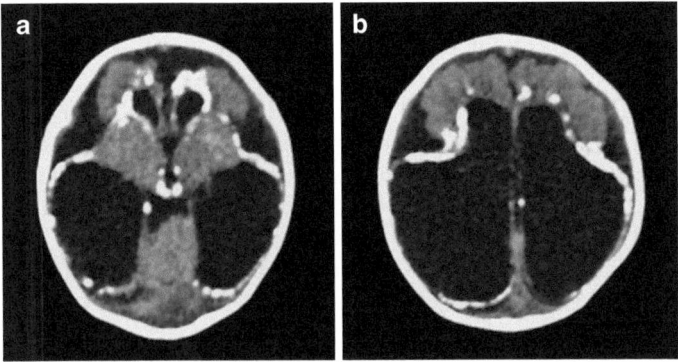

FIGURE 2 Unenhanced head CT of the child with congenital CMV infection shows (**a**) periventricular, cortical, thalamic and basal ganglia coarse calcifications, (**b**) decreased brain volume, extra-axial enlargement, and hydrocephalus. Image courtesy of Doctor Vanessa van der Linden

Calcification is present in 30–90% of cases and is frequently periventricular, being punctiform, linear, and coarse, the latter being the most common. There may be calcifications in the basal ganglia and cortical and white matter, which are generally asymmetric. Other findings are abnormalities of the white matter, neuronal migration disorder, cerebral and cerebellar atrophy, cysts, and ventriculomegaly [6, 10–13] (Fig. 2).

Toxoplasmosis

Toxoplasma gondii is an intracellular parasitic protozoan that affects approximately one third of the population worldwide. The percentage of subclinical congenital toxoplasmosis is 85%. The rate of infection of the fetus and the severity of the disease are inversely related. Early infection of the pregnant woman (first or second trimester) has a low possibility of fetal infection (9–27%) but, if it occurs, can cause severe congenital toxoplasmosis, intrauterine fetal death, and spontaneous abortion. On the other hand, late mother infection (third trimester) has a higher chance of fetal infection (up to 60% in the third trimester), but the consequences for the fetus are less likely, and if they exist, they are not as serious.

The neonatal clinical manifestations of toxoplasmosis can be varied including hydrocephalus, diffuse intracranial calcification, chorioretinitis, strabismus, blindness, epilepsy, mental or psychomotor retardation, petechia due to thrombocytopenia, and anemia [8].

The main alterations of the image are hydrocephalus and periventricular, cortical, thalamic, and basal ganglia calcification [14]. Calcifications are present in 50–80% of the cases [15] (Figs. 3 and 4).

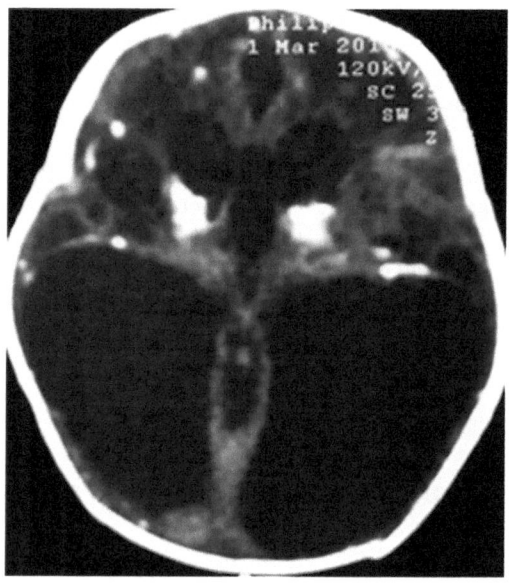

FIGURE 3 Unenhanced head CT in newborn with congenital toxoplasmosis shows hydrocephalus; periventricular, cortical, basal ganglia, and thalamic calcifications; and encephalomalacic changes. Image courtesy of Doctor Vanessa van der Linden

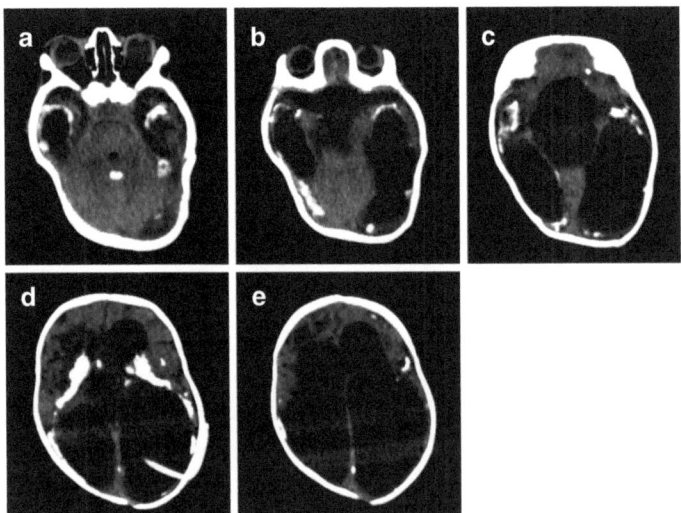

FIGURE 4 Unenhanced head CT in newborn with congenital toxo-plasmosis shows cranial deformity; non-communicant hydrocepha-lus; periventricular, cortical, basal ganglia, and thalamic calcifications (**a**, **b**, **c**, **d**) and encephalomalacic changes in the left frontal lobe (**d**, **e**). There is a ventricular derivation catheter in posterior horn of the left ventricle (**d**). Image courtesy of Doctor Eduardo Just

Syphilis

Congenital syphilis can be acquired from a mother infected by transplacental transmission of *T. pallidum* at any time during pregnancy. The Pan American Health Organization (PAHO) estimates that 330,000 pregnant women in Latin American and Caribbean countries with positive labora-tory results for syphilis are not treated at prenatal visits. Although the disease stage is a determining factor, it is estimated that two-thirds of maternal syphilis cases during pregnancy cause congenital syphilis or miscarriage. One hundred ten thousand children a year are born with con-genital syphilis, and a similar number of pregnancies end in fetal loss. In 20% of cases of maternal syphilis, births are premature.

Intrauterine infection can occur at any gestational age. However, pathological changes in fetal tissues cannot be detected before the 18th week of gestation. This is probably due to the lack of immune/inflammatory response of the fetus [8].

The main abnormalities are hepatosplenomegaly, placentomegaly, fetal ascites, hydrops fetalis, and long bone disorders [16].

The main neuroimaging findings are hydrocephalus, leptomeningeal enhancement, and stroke [7].

Rubella

Rubella remains endemic in many parts of the world. The rubella virus has a teratogenic action causing severe congenital malformations.

Rubella is a contagious disease that is transmitted through the respiratory system through the secretions of the nose or throat. Viremia occurs 5–7 days after contact during which the virus can be transmitted from the mother to the fetus by hematogenic transplacental spread. When infection occurs in the first trimester, the consequences are severe and can cause miscarriages, fetal death/stillbirths, and several congenital anomalies known as congenital rubella syndrome (CRS).

Before the vaccine was available, approximately every 6–9 years, there was a rubella epidemic. With the introduction of the vaccine in many countries, the incidence of rubella and congenital rubella syndrome has decreased significantly. In some countries, such as Finland, the United States, and Cuba, rubella and congenital rubella syndrome have been eradicated. The success of vaccination programs in some industrialized countries meant that the proportion of women of reproductive age susceptible to rubella was reduced to

2–3%. However, in developing countries that have not introduced the rubella vaccine into their national immunization programs, the proportion of susceptible women can range from less than 10% to more than 25%. In countries where only women are vaccinated, men remain susceptible and a possible source of infection for pregnant women susceptible to rubella. Rubella is a vaccine-preventable disease. It is assumed that the immunization provided by the vaccine lasts a lifetime, with data confirming immunity for more than 16 years [8].

Congenital rubella is usually associated with meningitis, ventriculitis, and subsequent ventriculomegaly. Calcifications are commonly of periventricular location in the basal ganglia and brain stem [7, 14].

Herpes Simplex Virus (HSV)

Primary HSV infection in the first trimester of pregnancy is associated with an increased risk of early miscarriage, but there is no pattern of fetal abnormalities in continuing gestations.

The great majority of neonatal cases occur by contagion through contact in the passage through the birth canal. Transplacental infection or ascending infection is less frequent. The risk of transmission is low (<1%) in women who have been infected in the first half of pregnancy or with a history of recurrent genital herpes lesions.

Although rare, neonatal infection can cause a neonatal syndrome due to disseminated HSV or encephalitis, which can be fatal or produce permanent sequelae [8].

Early intrauterine infection has a teratogenic neurotropic potential and is associated with intracranial (mainly periventricular) calcifications, microphthalmia, and retinal dysplasia [17].

The main imaging findings are extensive brain destruction, multicystic encephalomalacia, and scattered calcifications [7, 14].

Varicella Zoster Virus (VZV)

Chicken pox is a highly contagious childhood disease caused by the varicella zoster virus (VZV). The virus is spread by contact with the lesions or by the oropharynx of infected people. After the first infection, the VZV can remain latent in the dorsal root ganglion cells of the spinal cord and can be reactivated, which can happen many decades later in the form of herpes zoster. When primary infection occurs during pregnancy, consideration should be given to the possible consequences for the mother and the child. Chicken pox in pregnancy can cause severe disease in the mother, fetal death, and rarely congenital varicella syndrome (CVS). Maternal infection near the time of delivery has a risk of severe neonatal disease.

When chicken pox sets in the first 20 weeks of gestation, it can cause fetal loss (2.6%); after the 20th week, intrauterine death (0.7%). Chicken pox in pregnant women can also cause asymptomatic fetal infection. The proportion of babies with asymptomatic intrauterine infection may rise from 5% to 10% in the first and second quarters to approximately 25% around the 36th week of pregnancy.

CVS can occur when babies are born to infected mothers in the first half of pregnancy (during the first 20 weeks of gestation), and their manifestations may be low birth weight, cutaneous lesions, limbic hypoplasia, cerebral cortical atrophy, ventriculomegaly, chorioretinitis, cataract, and other ophthalmological complications. The risk of CVS is higher (2%) when the mother falls ill between the 13th and 20th week of pregnancy [8].

The main clinical abnormalities are in the skin, neurological, ocular, and skeletal [18, 19].

The main neuroimaging findings are hydrocephalus, cerebellar aplasia, polymicrogyria, and necrosis of the deep gray matter [7].

Human Immunodeficiency Virus (HIV)

The human immunodeficiency virus (HIV) is a retrovirus that infects the T cells of the immune system, causing a progressive reduction in its number and eventually AIDS— acquired immunodeficiency syndrome. HIV infection is a pandemic that affects more than 2.5 million children worldwide. Most children are infected during the peripartum period. It is estimated that without intervention in places without breastfeeding, approximately half of cases of mother-to-child transmission occur in the third trimester (from the 36th week) during labor or delivery [8].

The main abnormalities are intrauterine growth restriction, prominent forehead, and hypertelorism [20].

The main neuroimaging findings are meningoencephalitis, cerebral atrophy with consequent prominence of the subarachnoid space and the ventricles, and calcifications in vessels, basal ganglia, and subcortical white matter. The higher the HIV viral load, the more calcifications are seen on head CT [7].

References

1. Hazin AN, Poretti A, Di Cavalcanti Souza Cruz D, et al. Computed tomographic findings in microcephaly associated with Zika virus. N Engl J Med. 2016;374:2193–5.
2. de Fatima Vasco Aragao M, van der Linden V, Brainer-Lima AM, Coeli RR, Rocha MA, Sobral da Silva P, Durce Costa Gomes de Carvalho M, van der Linden A, Cesario de Holanda A, Valenca MM. Clinical features and neuroimaging (CT and MRI) findings in presumed Zika virus related congenital infection and microcephaly: retrospective case series study. BMJ. 2016;353:i1901.

3. Soares de Oliveira-Szejnfeld P, Levine D, Melo AS de O, et al. Congenital brain abnormalities and Zika virus: what the radiologist can expect to see prenatally and postnatally. Radiology. 2016;281:203–18.
4. Velho Barreto de Araújo T, Cunha Rodrigues L, Arraes de Alencar Ximenes R, et al. Association between Zika virus infection and microcephaly in Brazil, January to May, 2016: preliminary report of a case-control study. 2016. doi:10.1016/S1473-3099(16)30318-8.
5. Von der Hagen M, Pivarcsi M, Liebe J, von Bernuth H, Didonato N, Hennermann JB, Bührer C, Wieczorek D, Kaindl AM. Diagnostic approach to microcephaly in childhood: a two-center study and review of the literature. Dev Med Child Neurol. doi:10.1111/dmcn.12425.
6. Livingston JH, Stivaros S, Warren D, Crow YJ. Intracranial calcification in childhood: a review of aetiologies and recognizable phenotypes. Dev Med Child Neurol. 2014. doi:10.1111/dmcn.12359.
7. Raybaud C, Barkovich J. Pediatric neuroimaging. 5th ed. Philadelphia: Lippincott Williams & Wilkins, Wolters Kluwer; 2012.
8. CLAP/OPAS. Infecções Perinatais transmitidas de mãe para filho durante a gravidez. 2010.
9. Noyola DE, Demmler GJ, Nelson CT, et al. Early predictors of neurodevelopmental outcome in symptomatic congenital cytomegalovirus infection. J Pediatr. 2001;138:325–31.
10. Bale JF, Bray PF, Bell WE. Neuroradiographic abnormalities in congenital cytomegalovirus infection. Pediatr Neurol. 1985;1:42–7.
11. Boppana SB, Fowler KB, Vaid Y, Hedlund G, Stagno S, Britt WJ, Pass RF. Neuroradiographic findings in the newborn period and long-term outcome in children with symptomatic congenital cytomegalovirus infection. Pediatrics. 1997;99:409–14.
12. de Vries LS, Gunardi H, Barth PG, Bok LA, Verboon-Maciolek MA, Groenendaal F. The spectrum of cranial ultrasound and magnetic resonance imaging abnormalities in congenital cytomegalovirus infection. Neuropediatrics. 2004;35:113–9.
13. Fink KR, Thapa MM, Ishak GE, Pruthi S. Neuroimaging of pediatric central nervous system cytomegalovirus infection. Radiographics. 2010;30:1779–96.

14. Kiroğlu Y, Çalli C, Karabulut N, Öncel Ç. Intracranial calcifications on CT. Diagn Interv Radiol. 2010. doi:10.4261/1305-3825. DIR.2626-09.1.
15. Lago EG, Baldisserotto M, Hoefel Filho JR, Santiago D, Jungblut R. Agreement between ultrasonography and computed tomography in detecting intracranial calcifications in congenital toxoplasmosis. Clin Radiol. 2007;62:1004–11.
16. Reyna-Figueroa J, Esparza-Aguilar M, Hernández-Hernández L d C, Fernández-Canton S, Richardson-Lopez Collada VL. Congenital syphilis, a reemergent disease in Mexico: its epidemiology during the last 2 decades. Sex Transm Dis. 2011;38:798–801.
17. Dublin AB, Merten DF. Computed tomography in the evaluation of herpes simplex encephalitis. Radiology. 1977;125:133–4.
18. Smith CK, Arvin AM. Varicella in the fetus and newborn. Semin Fetal Neonatal Med. 2009;14:209–17.
19. Paryani SG, Arvin AM. Intrauterine infection with Varicella-Zoster virus after maternal varicella. N Engl J Med. 1986;314:1542–6.
20. AIDS embryopathy | Radiology Reference Article | Radiopaedia. org. https://radiopaedia.org/articles/aids-embryopathy.

Cases: Spectrum of Computed Tomography in Congenital Zika Syndrome

**Natacha Calheiros de Lima Petribu
and Maria de Fátima Viana Vasco Aragão**

Abbreviations

cm   Centimeters
CT   Computed tomography
GA   Gestational age
HC   Head circumference
SD   Standard deviation

―――――
N.C. de Lima Petribu (✉)
Hospital Barão de Lucena, Recife, PE, Brazil
e-mail: natachacalheiros@yahoo.com.br

M.F.V.V. Aragão
Centro Diagnóstico Multimagem, Mauricio de Nassau University,
Recife, PE, Brazil
e-mail: fatima.vascoaragao@gmail.com

M.F.V.V. Aragão (ed.), *Zika in Focus*,
DOI 10.1007/978-3-319-53643-9_8,
© Springer International Publishing AG 2017

Case 1. Axial unenhanced head CT and 3D reconstruction in a 5-day-old male neonate with confirmed congenital Zika virus infection, premature born at 31 weeks of pregnancy, with 810 g (appropriate for GA and sex) and HC of 23 cm   (below -3SD for GA and sex of the median INTERGROWTH-21 Standard), classified as severe microcephaly (Fig. 1).

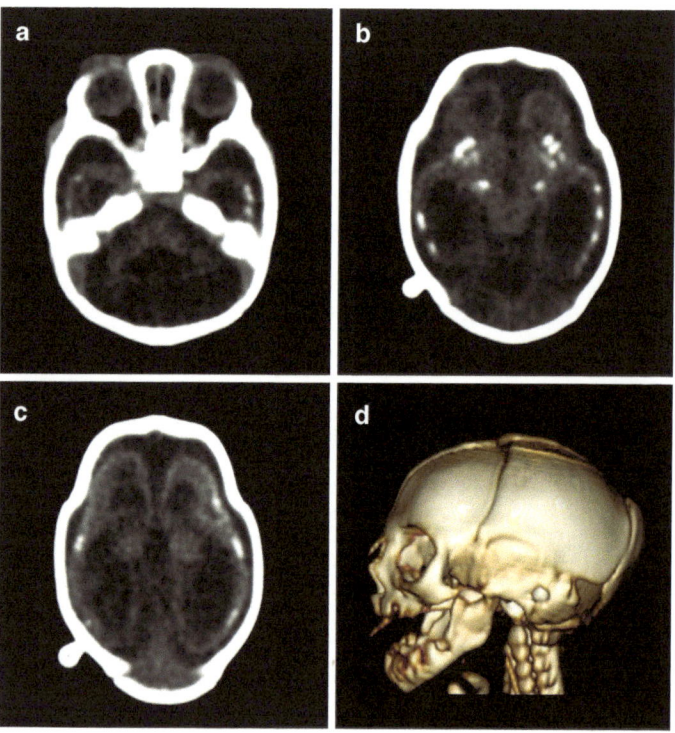

FIGURE 1 (**a**) Severe cerebellar hypoplasia, cisterna magna enlarged, dilatation of temporal horns of lateral ventricles and forth ventricle. (**b**) Colpocephaly, decreased brain volume, extra-axial enlargement, Malformation of cortical development, and coarse calcification in basal ganglia, thalamus, and cortico-subcortical junction. (**c**) Ventriculomegaly. (**d**) Prominent occipital bone

Case 2. Axial unenhanced head CT in a 5-day-old female neonate with confirmed congenital Zika virus infection, born at 37 weeks of pregnancy, with 2.700 g (appropriate for GA and sex) and HC of 32 cm, classified as normocephaly based on INTERGROWTH-21 Size at Birth Standards (Fig. 2).

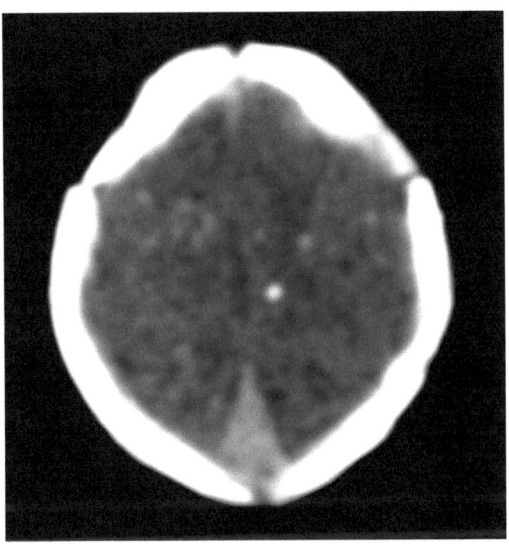

FIGURE 2 Punctiform calcification in the cortico-subcortical junction of frontal lobes

Case 3. Axial unenhanced head CT and 3D reconstruction in a 2-week-old male neonate with confirmed congenital Zika virus infection, born at 38 weeks of pregnancy, with 2.975 g (appropriate for GA and sex) and HC of 29 cm (below -3SD for GA and sex of the median INTERGROWTH-21 Standard), classified as severe microcephaly (Fig. 3).

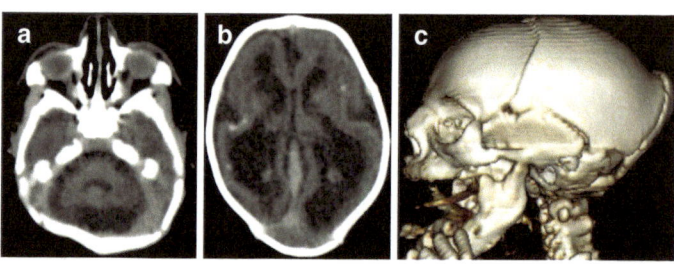

FIGURE 3 (**a**) Severe cerebellar hypoplasia associated with enlarged cisterna magna. (**b**) Ventriculomegaly, colpocephaly, decreased brain volume, extra-axial enlargement, Malformation of cortical development, and punctiform calcification in the cortico-subcortical junction of left frontal lobe and right temporal lobe. (**c**) Prominent occipital bone

Case 4. Axial unenhanced head CT and 3D reconstruction in a 3-week-old male neonate with confirmed congenital Zika virus infection late preterm born at 35 weeks of pregnancy, with 1.890 g (appropriate for GA and sex) and HC of 25 cm (below -3SD for GA and sex of the median INTERGROWTH-21 Standard), classified as severe microcephaly (Fig. 4).

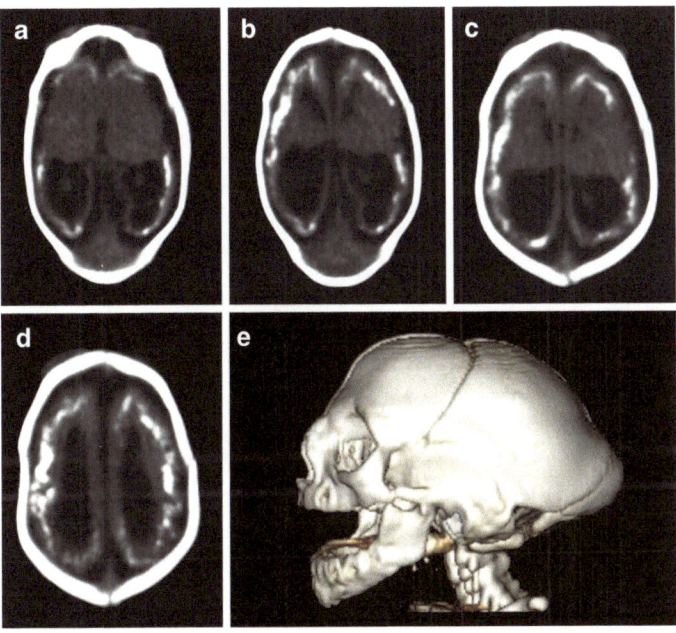

FIGURE 4 (**a–c**) Colpocephaly. (**a–d**) Decreased brain volume, extra-axial enlargement, malformation of cortical development, puncti-form, linear and coarse calcification in cortico-subcortical junction of all lobes. (**d**) Ventriculomegaly. (**e**) Prominent occipital bone

Case 5. Axial unenhanced head CT in a 1-day-old male neonate with confirmed congenital Zika virus infection, born at 40 weeks of pregnancy, with 3.450 g (appropriate for GA and sex) and HC of 32 cm, classified as normocephaly based on INTERGROWTH-21 Size at Birth Standards (Fig. 5).

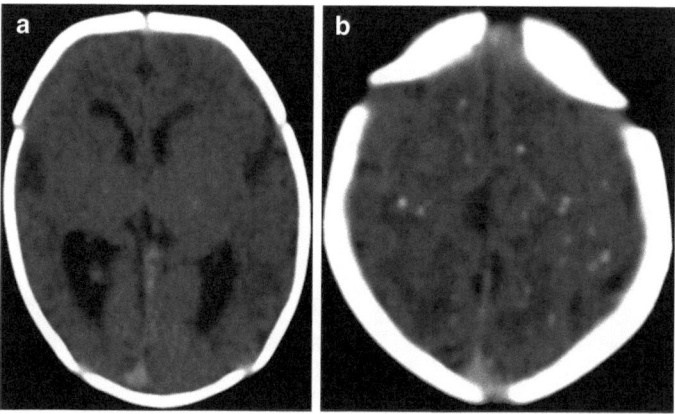

FIGURE 5 (**a**) Colpocephaly. (**b**) Punctiform calcification in the cortico-subcortical junction of frontal and parietal lobes

Case 6. Axial unenhanced head CT in a 3-week-old male neonate with confirmed congenital Zika virus infection, born at 39 weeks of pregnancy, with 3.840 g (appropriate for GA and sex) and HC of 33 cm classified as normocephaly based on INTERGROWTH-21 Size at Birth Standards (Fig. 6).

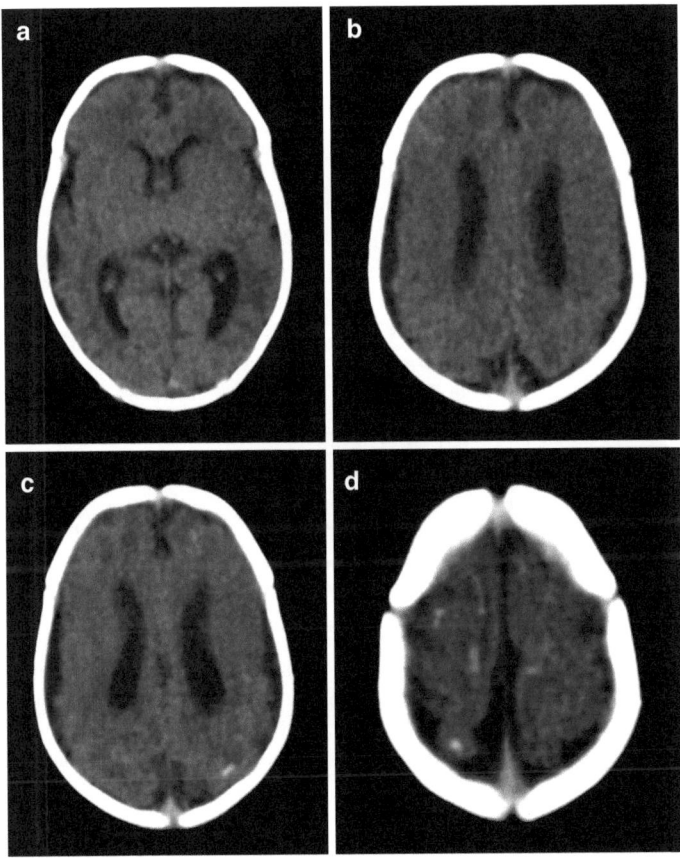

FIGURE 6 (**a**) Decreased brain volume and colpocephaly. (**b**, **c**) Ventriculomegaly and malformation of cortical development. (**c**, **d**) Punctiform calcification in the cortico-subcortical junction of frontal, parietal, and occipital lobes

Case 7. Axial unenhanced head CT and 3D reconstruction in a 2-week-old female neonate with confirmed syphilis and congenital Zika virus infection (coinfection), born at 38 weeks of pregnancy, with 2.700 g (appropriate for GA and sex) and HC of 30 cm (below -2SD for GA and sex of the median INTERGROWTH-21 Standard), classified as microcephaly (Fig. 7).

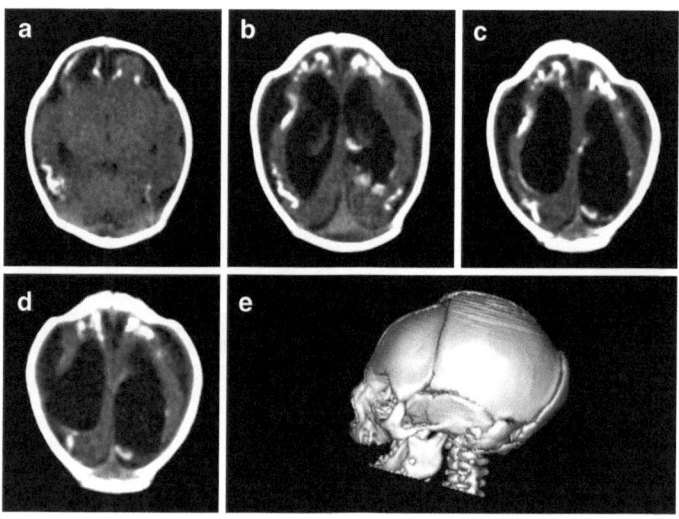

Figure 7 (**a**) Encephalomalacia in right frontal lobe. (**a–d**) Decreased brain volume, extra-axial enlargement, ventriculomegaly, malformation of cortical development, and coarse calcification in cortical, periventricular, and cortico-subcortical junction of all lobes. (**d**) Open-lip schizencephaly on right hemisphere. (**e**) Prominent occipital bone

Case 8. Axial unenhanced head CT and 3D reconstruction in a 3-day-old female neonate with confirmed congenital Zika virus infection, late preterm born at 36 weeks of pregnancy, with 2.650 g (appropriate for GA and sex) and HC of 28 cm (below -2SD for GA and sex of the median INTERGROWTH-21 Standard), classified as microcephaly (Fig. 8).

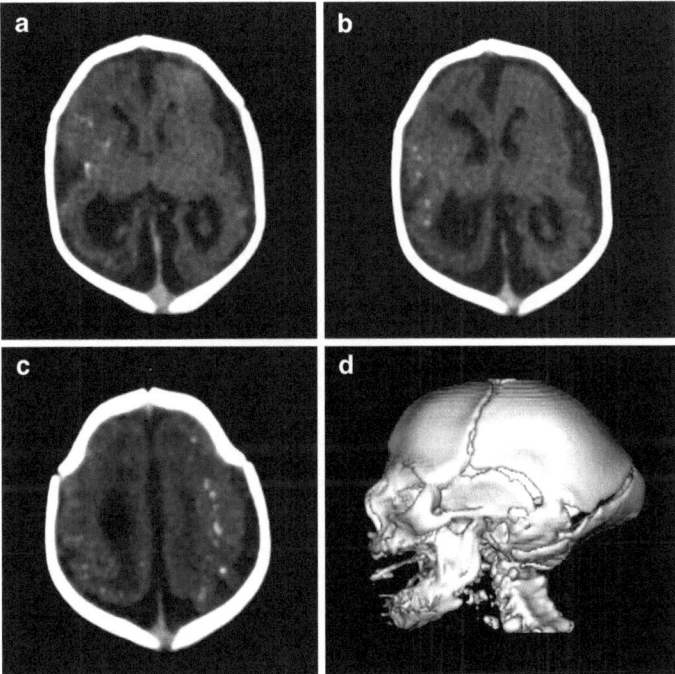

FIGURE 8 (**a**) Encephalomalacia in right occipital lobe. (**a**, **b**) Colpocephaly. (**a–c**) Decreased brain volume, extra-axial enlargement, ventriculomegaly, malformation of cortical development, and punctiform calcification in the cortico-subcortical junction of all lobes. (**d**) Prominent occipital bone

Case 9. Axial unenhanced head CT in a 1-day-old female neonate with confirmed congenital Zika virus infection, born at 40 weeks of pregnancy, with 2.880 g (appropriate for GA and sex) and HC of 28 cm (below -3SD for GA and sex of the median INTERGROWTH-21 Standard), classified as severe microcephaly (Fig. 9).

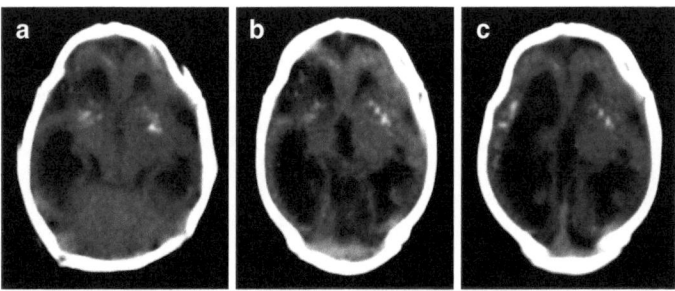

FIGURE 9 (**a–c**) Decreased brain volume, extra-axial enlargement, ventriculomegaly, malformation of cortical development, and punctiform calcification in the cortico-subcortical junction of frontal, parietal, and occipital lobes. (**a, b**) Colpocephaly and basal ganglia calcification

Case 10. Axial unenhanced head CT in a 3-day-old male neonate with confirmed congenital Zika virus infection, born at 38 weeks of pregnancy, with 2.975 g (appropriate for GA and sex) and HC of 32 cm classified as normo-cephaly based on INTERGROWTH-21 Size at Birth Standards (Fig. 10).

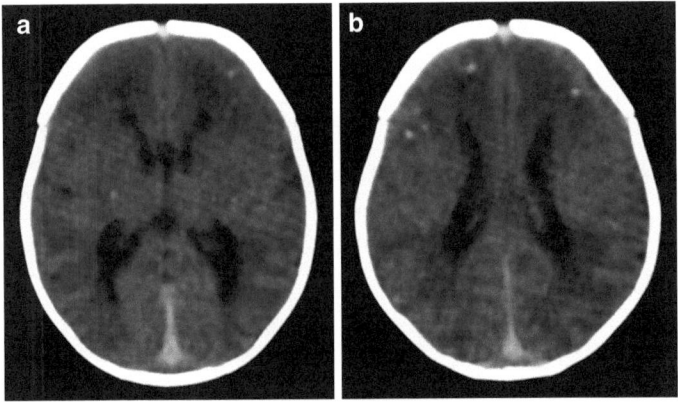

FIGURE 10 (**a**) Slight colpocephaly. (**a**, **b**) Punctiform calcification in the cortico-subcortical junction of frontal lobes

Case 11. Axial unenhanced head CT and 3D reconstruction in a 2-day-old female neonate with confirmed congenital Zika virus infection, born at 40 weeks of pregnancy, with 2.750 g (appropriate for GA and sex) and HC of 28 cm (below -3SD for GA and sex of the median INTERGROWTH-21 Standard), classified as severe microcephaly (Fig. 11).

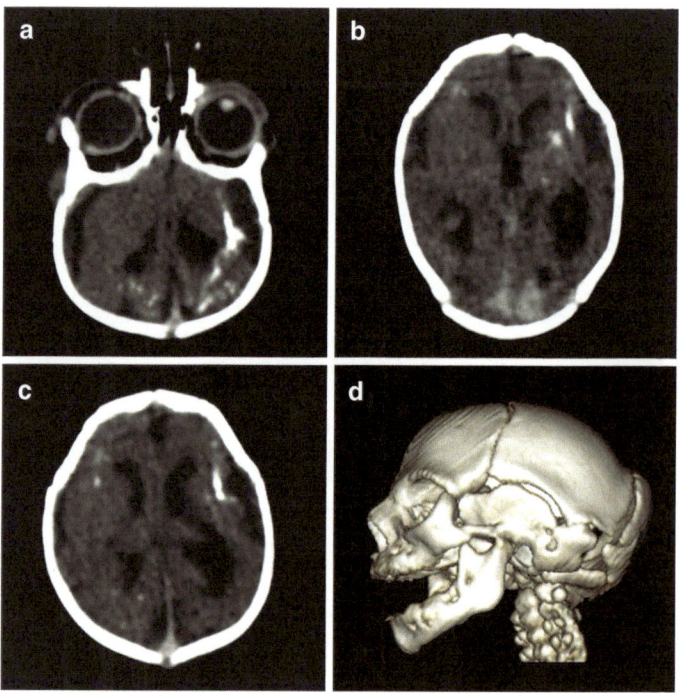

Figure 11 (**a–c**) Decreased brain volume, extra-axial enlargement, ventriculomegaly, colpocephaly, malformation of cortical development, punctiform and coarse calcification in the cortico-subcortical junction of frontal, parietal, temporal, and occipital lobes. (**d**) Prominent occipital bone

Case 12. Axial unenhanced head CT and 3D reconstruction in a 6-day-old male neonate with confirmed congenital Zika virus infection, born at 37 weeks of pregnancy, with 2.950 g (appropriate for GA and sex) and HC of 28 cm (below -3SD for GA and sex of the median INTERGROWTH-21 Standard), classified as severe microcephaly (Fig. 12).

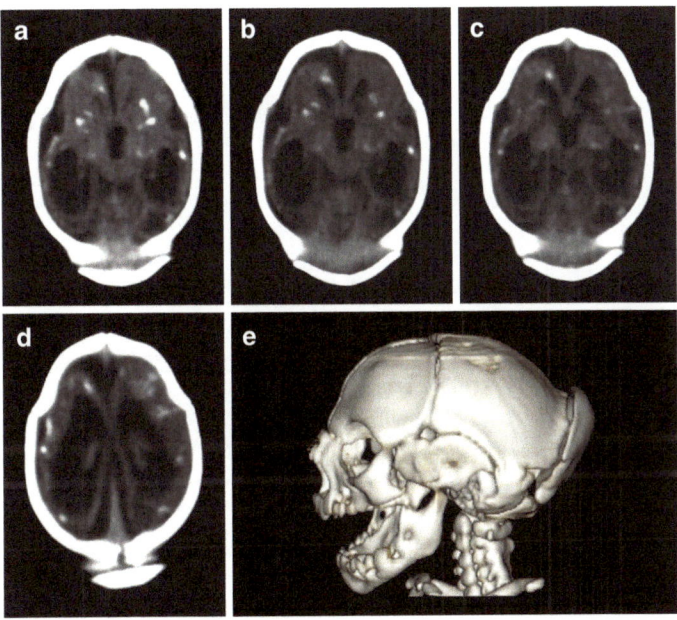

FIGURE 12 (**a–d**) Decreased brain volume, extra-axial enlargement, ventriculomegaly, malformation of cortical development, punctiform, linear and coarse calcification in the cortico-subcortical junction of frontal, parietal, temporal, and occipital lobes. (**a–c**) Colpocephaly. (**a, b**) Basal ganglia and thalamus calcification. (**e**) Prominent occipital bone

Case 13. Axial unenhanced head CT and 3D reconstruction in a 5-day-old male neonate with confirmed
congenital Zika virus infection, born at 39 weeks of pregnancy, with 2.600 g (appropriate for GA and sex) and HC
of 29 cm (below -3SD for GA and sex of the median
INTERGROWTH-21 Standard), classified as severe
microcephaly (Fig. 13).

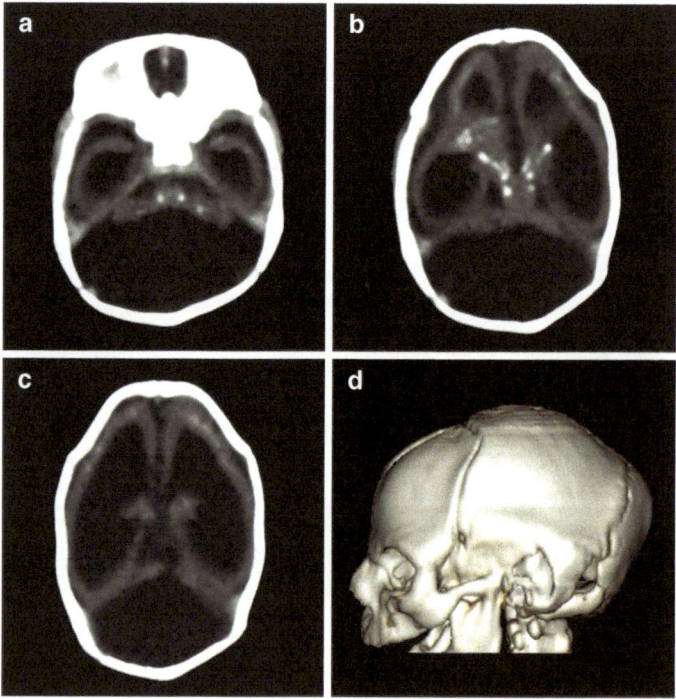

FIGURE 13 (**a**) Severe cerebellar hypoplasia associated to Dandy-
Walker complex, dilatation of temporal horn of lateral ventricle.
(**b**, **c**) Severe decreased brain volume, extra-axial enlargement, ventriculomegaly, malformation of cortical development, punctiform
and coarse calcification in brainstem, midbrain, basal ganglia, thalamus, and cortico-subcortical junction of frontal lobe. (**d**) Prominent
occipital bone

Case 14. Axial unenhanced head CT in a 5-day-old male neonate with confirmed congenital Zika virus infection, born at 37 weeks of pregnancy, with 2.970 g (appropriate for GA and sex) and HC of 29 cm (below -3SD for GA and sex of the median INTERGROWTH-21 Standard), classified as severe microcephaly (Fig. 14).

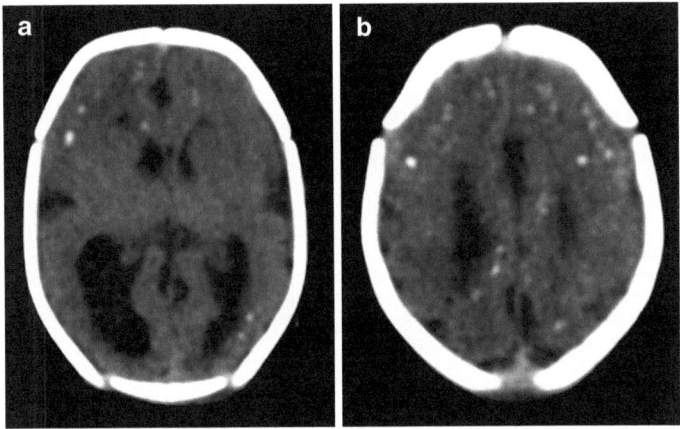

FIGURE 14 (**a**) Slight decreased brain volume, colpocephaly. (**a, b**) Punctiform calcification in the cortico-subcortical junction of frontal, parietal, temporal, and occipital lobes

Case 15. Axial unenhanced head CT and 3D reconstruction in a 2-week-old female neonate with confirmed
congenital Zika virus infection, born at 39 weeks of pregnancy, with 2.600 g (appropriate for GA and sex) and HC
of 29 cm (below -3SD for GA and sex of the median
INTERGROWTH-21 Standard), classified as severe
microcephaly (Fig. 15).

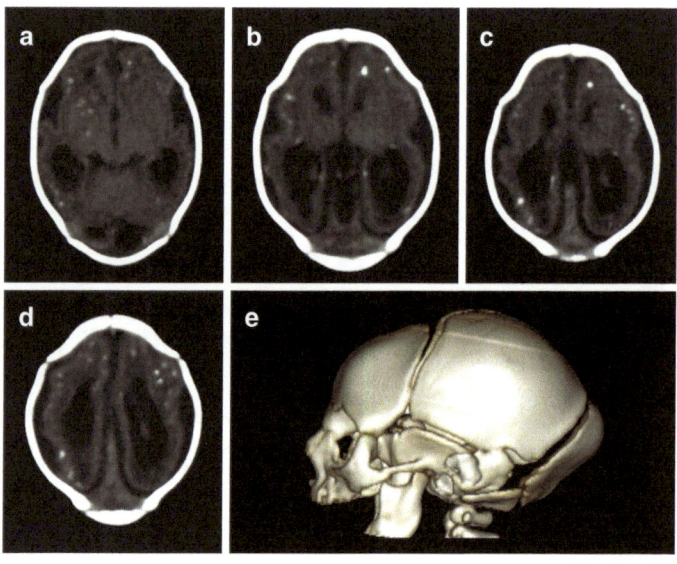

FIGURE 15 (**a–d**) Decreased brain volume, extra-axial enlargement,
ventriculomegaly, malformation of cortical development, and predominant punctiform calcification in the cortico-subcortical junction of frontal, parietal, temporal, and occipital lobes. (**a–c**)
Colpocephaly. (**e**) Prominent occipital bone

Case 16. Axial unenhanced head CT and 3D reconstruction in a 1-week-old female neonate with confirmed syphilis and congenital Zika virus infection (coinfection), born at 38 weeks of pregnancy, with 2.600 g (appropriate for GA and sex) and HC of 29 cm (below -3SD for GA and sex of the median INTERGROWTH-21 Standard), classified as severe microcephaly (Fig. 16).

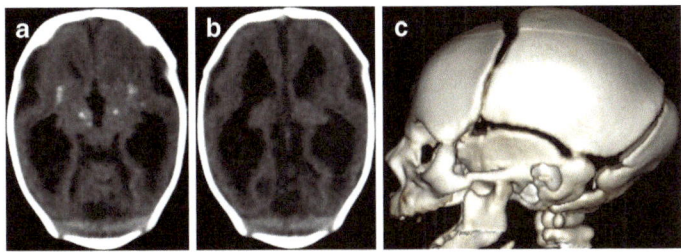

FIGURE 16 (**a, b**) Decreased brain volume, extra-axial enlargement, ventriculomegaly, colpocephaly, malformation of cortical development, punctiform calcification of basal ganglia and thalamus, and thin punctiform calcification in the cortico-subcortical junction of left frontal lobe. (**c**) Prominent occipital bone

Case 17. Surprising reduction of calcifications identified during follow-up of an infant with confirmed congenital Zika virus infection, one year after birth (Fig. 17).

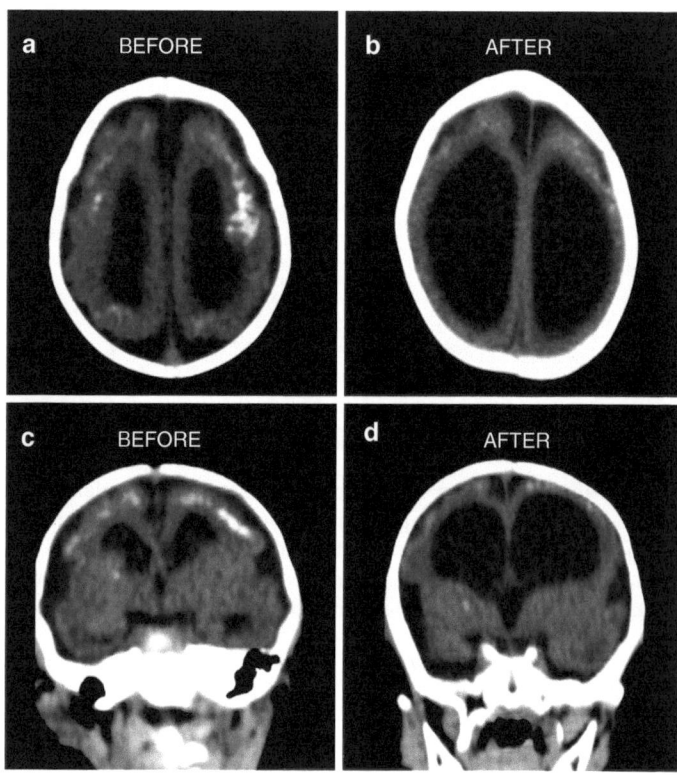

FIGURE 17 Axial head CT of an infant with confirmed congenital Zika syndrome (**a**) at birth and (**b**) after 1 year of birth, which shows reduction in the number and size of calcifications located in cortico-subcortical junction of frontal lobes and disappearance in the right parietal lobe. Coronal reconstruction head CT of the same infant (**c**) at birth and (**d**) after 1 year of birth, which shows reduction in the number and size of calcifications located in the cortico-subcortical junction and right basal ganglia. Some calcification disappeared. We can also note that in this case progression of cerebral atrophy and ventricular dilatation occurs

Cases: Spectrum of Magnetic Resonance Imaging in Congenital Zika Syndrome

Alessandra Mertens Brainer-Lima,
Maria de Fátima Viana Vasco Aragão,
and Arthur Cesário de Holanda

Abbreviations

cm	Centimeters
FFE	Fast field echo
FIESTA	Fast imaging employing steady-state acquisition
FLAIR	Fluid-attenuated inversion recovery
GA	Gestational age
HC	Head circumference
MRI	Magnetic resonance image
SWI	Susceptibility weight imaging

A.M. Brainer-Lima, M.D., M.Sc. (✉)
University of Pernambuco — PROCAPE, Recife, PE, Brazil
e-mail: mertensbrainer@yahoo.com.br

M.F.V.V. Aragão, M.D., Ph.D.
Centro Diagnóstico Multimagem, Mauricio de Nassau University,
Recife, PE, Brazil
e-mail: fatima.vascoaragao@gmail.com

A.C. Holanda
Universidade Federal de Pernambuco, Recife, PE, Brazil
e-mail: arthur.c.holanda@gmail.com

M.F.V.V. Aragão (ed.), *Zika in Focus*,
DOI 10.1007/978-3-319-53643-9_9,
© Springer International Publishing AG 2017

Case 1. Brain MRI of a 4-month-old female, confirmed congenital Zika virus infection, who was prematurely born at 35 weeks of pregnancy, with a HC of 29.5 cm. The mother reported rash at 2 months of pregnancy (Fig. 1).

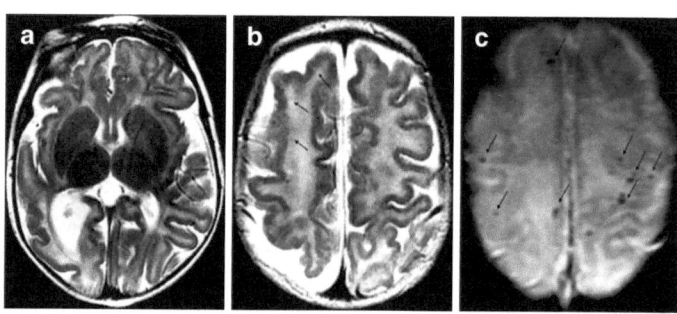

FIGURE I Axial T2-weighted images (**a**, **b**) show slight asymmetry between cerebral hemispheres, with volume decrease on the right occipitotemporal lobes and enlargement of the right ventricle. There is asymmetry of the gyral pattern, thickened right frontal lobe with irregularities of the inner cortical surface, related to polymicrogyria (*black arrow*). Axial SWI (**c**) shows multiple foci of low signal intensity consistent with calcifications (*arrows*). Reproduced with permission from de Fátima Vasco Aragão et al. [1]

Case 2. Brain MRI of a 1-month-old male neonate, confirmed congenital Zika virus infection, who was born at 37 weeks of pregnancy, with a HC of 27 cm at birth. The mother reported rash at 3 months of pregnancy (Fig. 2).

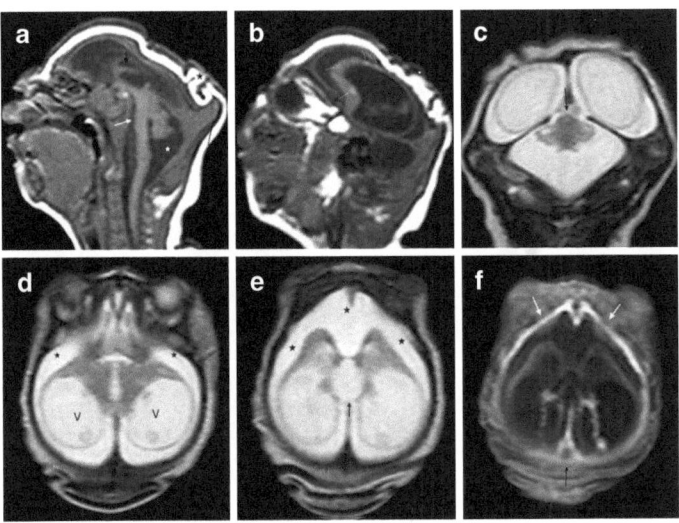

FIGURE 2 Severe microcephaly. Sagittal T1-weighted image (**a**) shows severe craniofacial disproportion, hypogenetic corpus callosum (*short black arrow*), and brainstem (*white arrow*) and cerebellum hypoplasia. In addition, there are enlarged cisterna magna (*white star*), redundant scalp skin (*black star*), and prominent occipital protuberance (*long black arrow*). Observe the dystrophic calcifications characterized by small hyperintense foci on parasagittal T1-weighted image (**b**) in the frontal lobe (*white arrows*), and extremely simplified gyral pattern. Coronal and axial T2-weighted images (**c**, **d**, and **e**) show cerebellum hypoplasia (*black arrow*) and severe ventriculomegaly, mainly at the posterior horn and ventricular atrium (*V*). Note the bulging walls of the ventricle, the upward dilated third ventricle (*black arrow*), and enlargement of the subarachnoid space (*stars*). Axial T1-weighted image fat suppression post-contrast (**f**) shows probable sagittal sinus thrombosis, thickness, and enhancement, secondary to pachymeningitis or to the thrombosis (*black arrows*). Reproduced with permission from de Fátima Vasco Aragão et al. [1]

Case 3. Brain MRI of a 1-month-old female neonate, probable congenital Zika virus infection, who was born at 40 weeks of pregnancy with a HC of 30 cm. The mother reported rash at 2 months of pregnancy (Fig. 3).

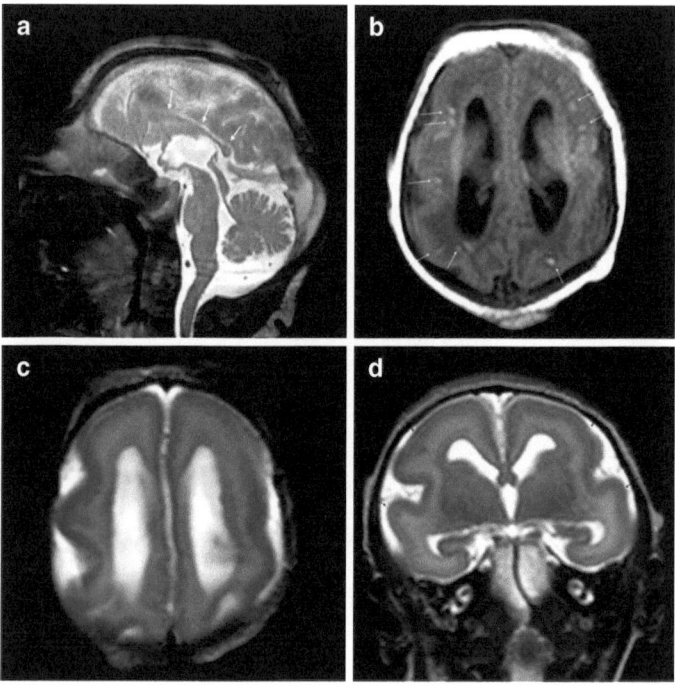

FIGURE 3 Brain MRI of infant with microlissencephaly. Sagittal T2-weighted image (**a**) shows corpus callosum hypoplasia and enlarged cisterna magna (*stars*). Axial T1 SE (**b**) shows multiple foci of calcifications with hyper signal intensity predominantly in the cortical and subcortical junction. In addition, note ventriculomegaly. Almost completely smooth cerebral surface with a diffusely thick cortex on axial (**c**) and coronal T2-weighted images (**d**). Reproduced with permission from de Fátima Vasco Aragão et al. [1]

Case 4: Brain MRI of a 3-month-old female, confirmed congenital Zika virus infection, who was born at 37 weeks of pregnancy with a HC of 30 cm at birth. The mother reported rash at 3 months of pregnancy (Fig. 4).

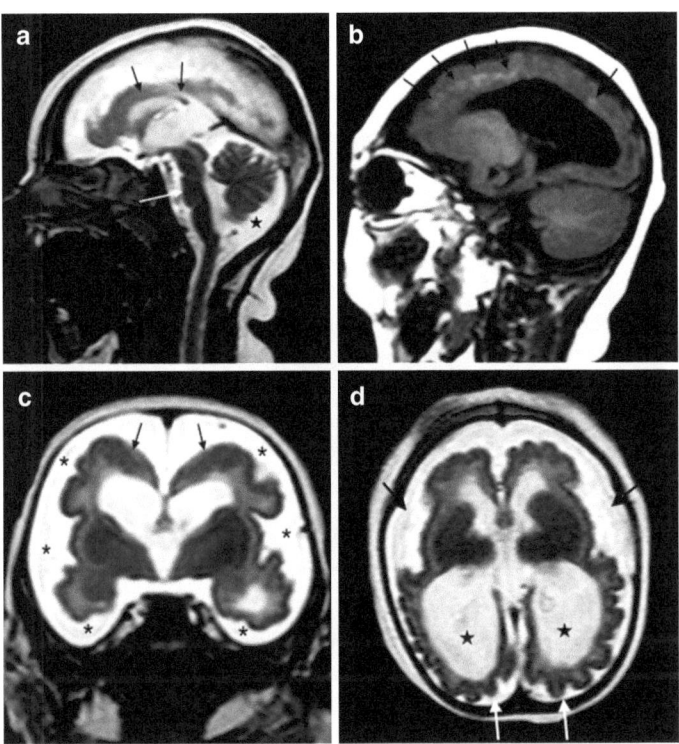

FIGURE 4 Sagittal T2-weighted image (**a**) shows hypogenesis of the corpus callosum (*black arrows*), enlarged cisterna magna (*star*), and pons hypoplasia (*white arrow*). Parasagittal T1 shows many small foci of high signal intensity in the cortical-subcortical junction and periventricular region (*black arrows*) consistent with calcifications. Coronal T2-weighted image (**c**) shows hyposulcation and simplified gyral pattern, bilaterally increased cortical thickness in the pachygyric frontal lobes (*black arrows*), and enlargement of subarachnoid space (*stars*). Axial T2-weighted image (**d**) shows posterior simplified gyral pattern (*white arrows*), ventriculomegaly (*stars*), wildly open Sylvian fissure, as well as enlargement of subarachnoid space (*black arrows*). Reproduced with permission from de Fátima Vasco Aragão et al. [1]

Case 5. Brain and spinal cord MRI of a 5-month-old male with reduced thoracic spinal cord thickness, probable congenital Zika virus infection, who was born at 36 weeks of pregnancy with a HC of 31.5 cm. The mother reported rash at 4 months of pregnancy (Fig. 5). There is no arthrogryposis, but the lesions are in the spectrum of Zika-related spinal cord damage [3].

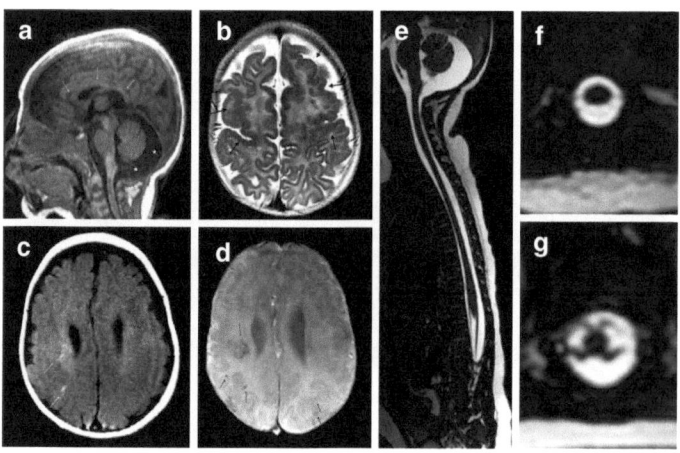

FIGURE 5 Sagittal T1-weighted image (**a**) shows hypoplasic corpus callosum (*white arrow*) and enlarged cisterna magna (*stars*). Axial T2-weighted image (**b**) shows bilateral frontal polymicrogyria (*black arrow*) with irregular cortico-subcortical interface. Linear and punctiform foci of T1 shortening on axial image (*white arrows* in **c**) and multiple foci of T2 hypointensity in subcortical frontal white matter on axial SWI (*black arrows* in **d**), which correspond to calcifications. Reproduced with permission from de Fátima Viana Vasco Aragão et al. [1]. Sagittal T2-weighted FIESTA (**e**) shows normal spinal cord. Axial reconstruction of T2-weighted FIESTA (**f**, **g**) shows normal conus medullaris and moderate reduction in thickness of the conus medullaris ventral roots compared with dorsal roots

Case 6. Brain and spinal cord MRI of a 3-month-old female with arthrogryposis, confirmed congenital Zika virus infection, who was born at 38 weeks of pregnancy with a HC of 30 cm at birth. The mother reported rash at 2 months of pregnancy (Fig. 6).

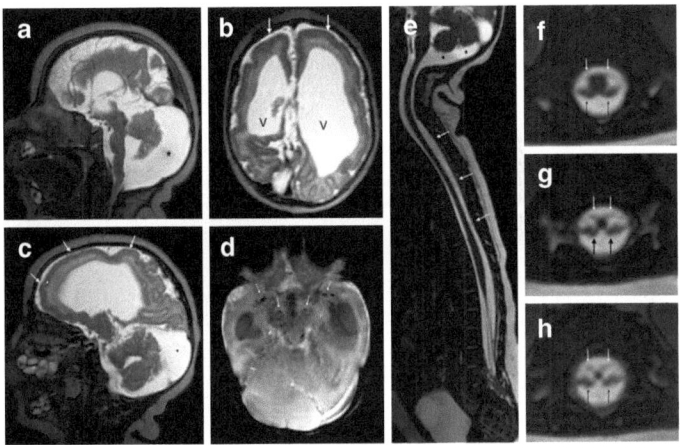

FIGURE 6 Brain and spinal cord MRI of an infant with arthrogryposis. Sagittal T2-weighted image (**a**) shows hypogenesis of corpus callosum (*black arrows*), enlarged cisterna magna (*star*), and enlarged fourth ventricle and pons hypoplasia. Axial T2-weighted imaging (**b**) shows pachygyria in frontal lobes (*white arrows*) and severe ventriculomegaly (*V*), mainly at posterior part of lateral ventricles. Note the posterior simplified gyral pattern against the pachygyric frontal appearance of the cortex in the frontal lobe (*white arrows*) (**c**). Axial susceptibility weighted image (**d**) shows some hypointense small dystrophic calcifications (*white arrows*) in midbrain and basal ganglia. The calcifications in the junction between cortical and subcortical white matter are not shown. Sagittal T2-weighted FIESTA (**e**) shows apparently reduced spinal cord thickness (*white arrows*) and enlarged cisterna magna (*stars*). Axial reconstruction of T2-weighted FIESTA (**f**, **g**, and **h**) shows reduction of medullary cone ventral roots (*white arrows*) compared with dorsal roots (*black arrows*). Reproduced with permission from Van der Linden et al. [2]

Case 7. Brain MRI of a 5-month-old female, probable congenital Zika virus infection, was born at 40 weeks of pregnancy with a HC of 30 cm at birth. The mother reported no rash during pregnancy (Fig. 7).

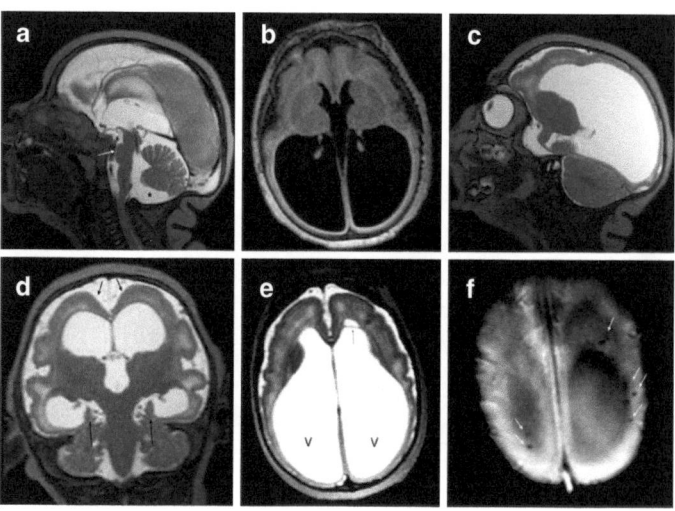

FIGURE 7 Sagittal T2-weighted image (**a**) shows hypogenetic corpus callosum, brainstem hypoplasia (*white arrow*), and enlarged cisterna magna (*star*). Axial T1-weighted image (**b**) shows severe ventriculomegaly, mainly at the posterior horn and ventricular atrium. Note the bulging walls of the ventricle. Parasagittal and coronal T2-weighted images (**c, d**) show hyposulcation and extremely simplified gyral pattern posteriorly (**c**), bilaterally increased cortical thickness in the pachygyric frontal lobes (*short black arrows*), and, additionally, malrotation of the hippocampus (*long black arrows*). Axial T2-weighted image (**e**) demonstrates slight septation (*black arrow*) inside the frontal horn of the left ventricle (*V*). Multiple foci of low signal intensity calcifications are seen on the axial SWI (**f**)

Case 8. Brain MRI of a 5-month-old male, confirmed congenital Zika virus infection, who was born at 39 weeks of pregnancy with a HC of 31 cm at birth. The mother reported rash at 3 months of pregnancy (Fig. 8).

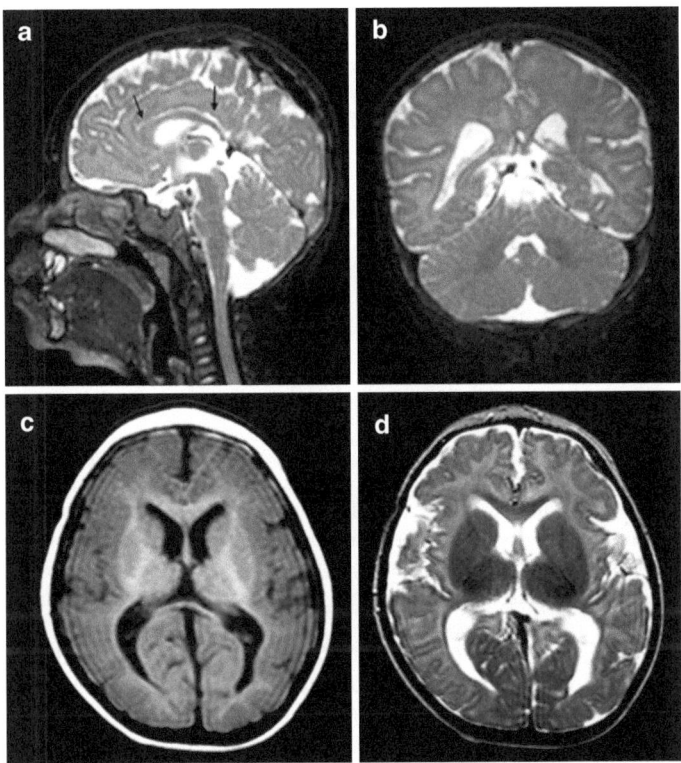

FIGURE 8 Sagittal T2-weighted image (**a**) shows slightly hypogenetic corpus callosum and enlarged cisterna magna. Coronal T2-weighted image (**b**) demonstrates normal gyral architecture and cerebellum. Axial T1- (**c**) and axial T2-weighted images (**d**) show mild ventriculomegaly and normal myelination pattern. Note the temporal subarachnoid space mildly enlarged on T2 (**d**). No calcifications nor definitive malformations of cortical development were detected

Case 9. Brain and spinal cord MRI of a 4-month-old female with arthrogryposis, confirmed congenital Zika virus infection, who was born at 37 weeks of pregnancy with a HC of 29 cm at birth. The mother reported no rash during pregnancy (Fig. 9).

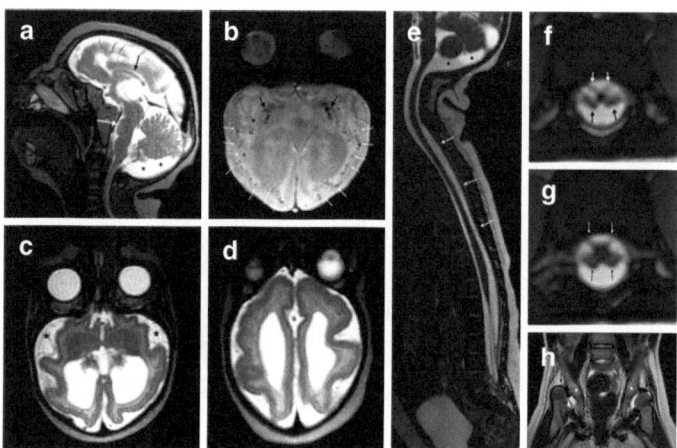

FIGURE 9 Microcephaly and arthrogryposis. Sagittal T2-weighted (**a**) image shows hypoplasic corpus callosum, pons hypoplasia, and enlarged cisterna magna (*stars*). Axial SWI (**b**) shows multiple foci of low signal intensity in the basal ganglia (*black arrows*) and in the cortical and subcortical white matter junction (*white arrows*). Axial T2-weighted images (**c, d**) show enlargement of subarachnoid space (*stars*), posterior simplified gyral pattern, and pachygyria in frontal lobes, as well as ventriculomegaly. Sagittal T2-weighted FIESTA (**e**) shows reduction of the entire spinal cord thickness (*short arrows*). Enlarged cisterna magna is seen (*stars*). Axial reconstruction of T2-weighted FIESTA (**f, g**) shows severe reduction of medullary cone ventral roots (*white arrows*) compared with dorsal roots (*black arrows*). Congenital bilateral hip luxation is seen on coronal T2-weighted image (**h**)

Case 10. Brain MRI of an almost 2-month-old female, confirmed congenital Zika virus infection, who was born at 38 weeks of pregnancy with a HC of 33 cm and 3520 g at birth. The mother reported rash during the end of third month of pregnancy (Fig. 10).

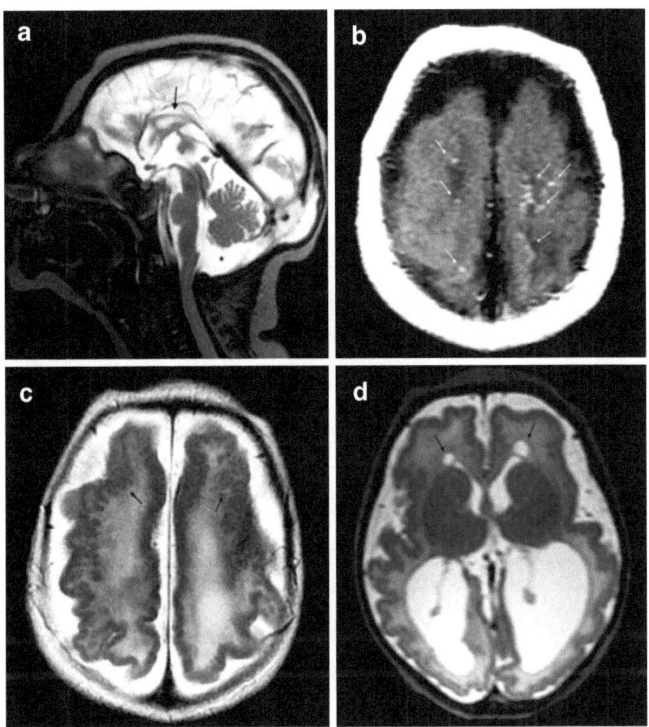

FIGURE 10 Sagittal T2-weighted image (**a**) shows hypogenetic corpus callosum (*arrow*), mild pons hypoplasia, and enlarged cisterna magna (*star*). Axial T1-weighted image (**b**) shows many small foci of high signal intensity in the junction between cortical and subcortical white matter (*arrows*). Axial T2-weighted image (**c**) shows thickened cortex with irregularities of the inner and outer cortical surfaces, predominantly in frontal lobes, and focal areas posteriotly, in left parietal and occipital lobes. However, the parietal, occipital and temporal lobes have predominantly thick cortex with smooth inner and outer surfaces, looking like pachygyria. Axial T2-weighted image (**d**) demonstrates slight subventricular cysts/ventricular septation in frontal horns of the ventricles (*black arrows*). In addition, note severe ventriculomegaly, mainly at the posterior horn and ventricular atrium

Case 11. Brain MRI of a male infant, with 2 months and 15 months of age, and who was born at 38 weeks of pregnancy with a HC of 29.5 cm at birth, confirmed congenital Zika virus infection. The mother reported rash during 3 months of pregnancy (Fig. 11).

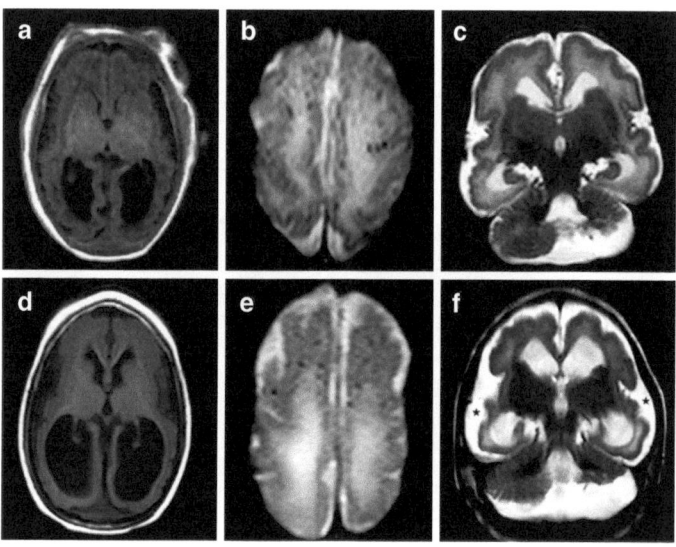

FIGURE 11 (*First row*, **a–c**) First MRI of a male infant with almost 2 months of age. Axial T1-weighted image (**a**) shows ventriculomegaly, mainly at the posterior horn and ventricular atrium of the ventricle. Multiple foci of low signal intensity on FFE (**b**) predominantly in cortical-subcortical junction, consistent with calcifications. Coronal T2-weighted image (**c**) shows slight enlargement of temporal subarachnoid spaces. (*Second row*, **d–f**) Follow-up MRI of the same infant, with 15 months of age, showing evolution of the brain lesions, with reduced calcifications and brain thickness, when compared to the first MRI. Axial T1-weighted image (**c**) shows more severe ventriculomegaly, with greater dilatation of the whole supratentorial ventricular system, more evident at the temporal, posterior horn, and atrium of the ventricles. Notice that there are less foci of calcifications on FFE (**e**) than in the previous MRI (**b**). Increased subarachnoid spaces (*stars*) and reduction of brain thickness are well seen on coronal T2-weighted image (**f**)

References

1. de Fatima Vasco Aragao M, van der Linden V, Brainer-Lima AM, Coeli RR, Rocha MA, Sobral da Silva P, Durce Costa Gomes de Carvalho M, van der Linden A, Cesario de Holanda A, Valenca MM. Clinical features and neuroimaging (CT and MRI) findings in presumed Zika virus related congenital infection and microcephaly: retrospective case series study. BMJ. 2016;353:i1901.
2. 2. Linden V v d, Filho ELR, Lins OG, et al. Congenital Zika syndrome with arthrogryposis: retrospective case series study. BMJ. 2016;354:i3899.
3. de Fatima Vasco Aragão M, Brainer-Lima AM, Holanda AC, van der Linden V, Aragão LV, Silva Júnior MLM, Sarteschi C, Petribu NCL, van der Linden A, Valença MM. (IN PRESS) Spectrum of spinal cord, spinal roots and brain MRI abnormalities in congenital Zika syndrome with and without arthrogryposis. AJNR. 2017.

Histopathological Findings of Congenital Zika Syndrome

Arthur Cesário de Holanda
and Roberto José Vieira de Mello

The macroscopic brain abnormalities found in congenital Zika syndrome have been well characterized by imaging studies [1–5]. There are not as many cases evaluated by anatomopathological studies [6–12]. However, they were important in providing support for the initial association between Zika virus and microcephaly, as well as information in regard to how the virus acts on the brain. A brief comprehension of the major abnormalities found in macroscopic and microscopic evaluations of congenital Zika syndrome, therefore, helps to understand the findings in neuroimaging scans of infants with the syndrome.

The first reports with histopathological findings associated with Zika virus were published, in fact, before the neuroimaging features were known by the scientific and health communities. Since Mlakar and colleagues [9] reported the case of a fetus with microcephaly and other brain abnormalities, whose mother had been presumably infected by the virus during the first trimester of pregnancy, only other six studies have been published, yielding the analysis of four fetuses and eight babies [7–13].

A.C. Holanda (✉) • R.J.V. Mello, M.D., Ph.D.
Universidade Federal de Pernambuco, Recife, PE, Brazil
e-mail: arthur.c.holanda@gmail.com; rjvmello@gmail.com

M.F.V.V. Aragão (ed.), *Zika in Focus*,
DOI 10.1007/978-3-319-53643-9_10,
© Springer International Publishing AG 2017

According to our experience, cases of fetal death are very difficult to assess, due to the condition of complete destruction in which most of the brains are found. When death occurs shortly after birth, macroscopic and microscopic analysis can usually be carried out, even though the condition of many of the specimens is still not ideal. Direct viral damage, hypoxia, maceration, and difficulties in brain removal and fixation processes impair preservation of the fragile fetal specimens. Newborn specimens, in their turn, suffer less with these factors.

Macroscopic Findings

In preserved brain specimens, pathologic findings correspond to the abnormalities identified in radiological studies, especially by magnetic resonance imaging (MRI). The overall macroscopic feature usually found is a brain decreased in size, with severely underdeveloped lobes and abnormal gyration and sulcation, as seen in Figs. 1 and 2. Usually, sulci are shallow and their anatomical distribution is difficult to be determined, while gyri can be broad and few in number (pachygyria) or very small and numerous (polymicrogyria). Some cases present agyria (complete lissencephaly), in which no sulci or gyri can be identified and opercularization of the insula does not occur, as shown in Fig. 3. Brainstem and cerebellum hypoplasias are identified in some cases (Fig. 1b), but these structures can be well preserved, despite of the severe brain damage (Fig. 4a and b).

Brain slices reveal ventriculomegaly (Figs. 1c and 4c) and increased cortical thickness (Figs. 1c, 2b, 3b, and 4c). They also show that subcortical nuclei are not compromised (Fig. 4c).

In the literature, cases reported range from microcephaly with apparently normal gross anatomy [6] to complete lissencephaly. One case of alobar holoprosencephaly has been reported [8]. Both in fetal and postnatal autopsies, abnormalities such as malformations of cortical development (simplified gyral cortex, pachygyria, agyria, and polymicrogyria), cerebellum and brainstem hypoplasias, and ventriculomegaly

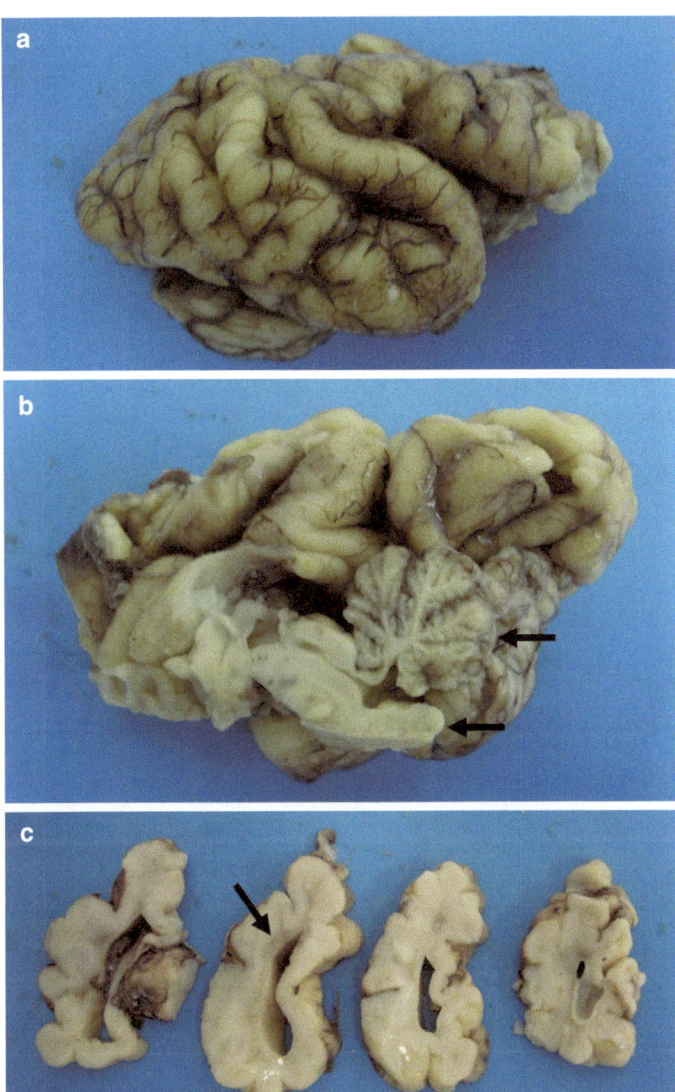

FIGURE 1 Lateral (**a**) and medial (**b**) views of a brain specimen with severe reduction in size and brainstem and cerebellum hypoplasias (*black arrows*). The cortex is pachygyric. Coronal brain slices (**c**) show enlarged ventricle, shallow sulci, and thick cortex, with much reduced white matter (*black arrow*)

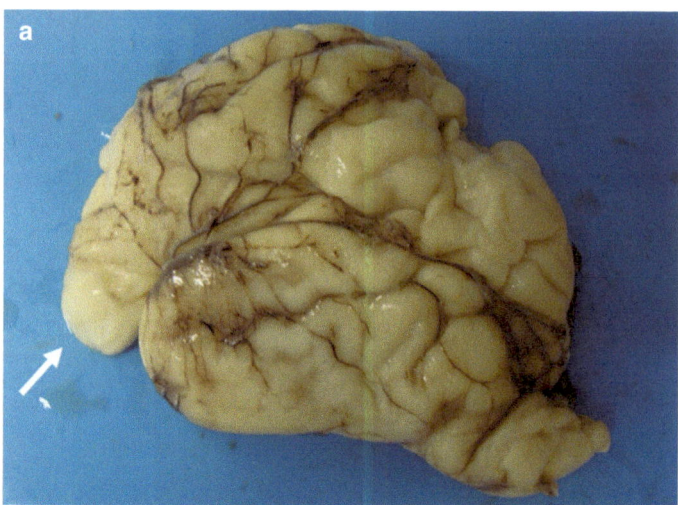

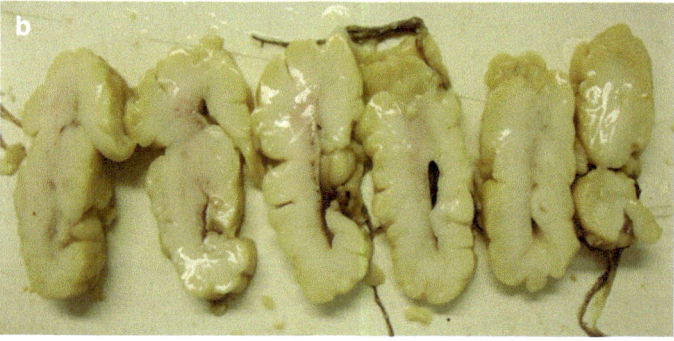

FIGURE 2 Lateral view (**a**) of a brain specimen with severe reduction in size, especially of the frontal lobes (*white arrow*). The cortex is pachygyric, with very few sulci identified. Coronal brain slices (**b**) show shallow sulci

have been reported [7–9, 12]. Subcortical nuclei were found to be well developed at gross inspection [9].

Other congenital abnormalities have also been described in some of the studies, as limb malformations and arthrogryposis, craniofacial malformations, intrauterine growth restriction, craniosynostosis, and pulmonary hypoplasia [7, 8, 10]. It is believed that these findings are not direct consequences of the

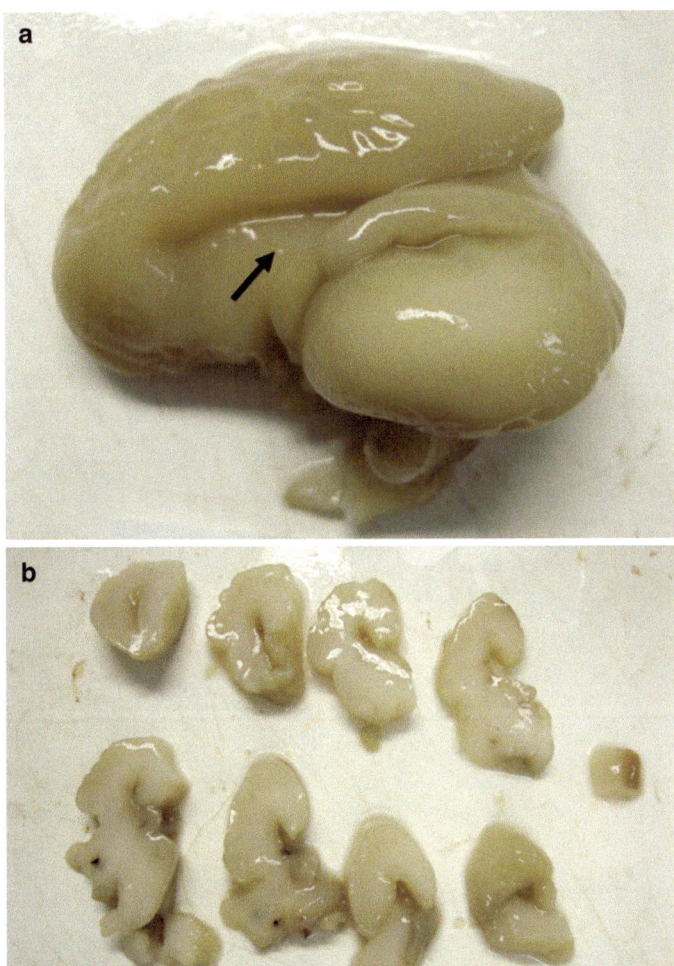

FIGURE 3 Lateral view (**a**) of a brain specimen with severe complete lissencephaly (agyria). There are no sulci and gyri, and the insula is exposed (*black arrow*). Coronal brain slices (**b**) show the smooth cortex, without sulci

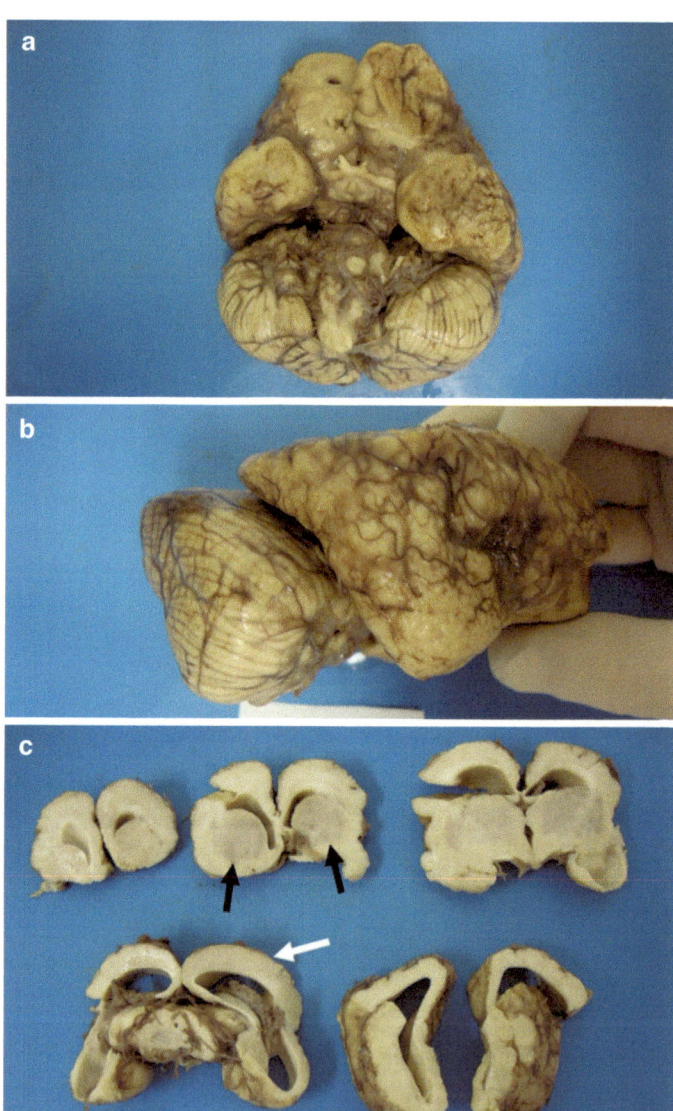

FIGURE 4 Ventral (**a**) and lateral (**b**) views of a brain specimen with severe reduction in size and agyria. The cerebellum is well developed, with almost the size as the brain. Coronal brain slices (**c**) show ventriculomegaly, smooth cortex with increased thickness (*white arrow*), and subcortical nuclei apparently preserved in size (*black arrows*)

viral cytopathic effects, being part of fetal akinesia deformation sequence [14]. However, in regard to arthrogryposis, other evidence suggest involvement of the virus [13, 15]; for example, it has been found in the spinal cord of these infants [13].

Microscopic Findings

At microscopy, the most prominent feature is an abnormal distribution of neuroblasts at the germinal matrix and at the neocortex, as shown in Fig. 5. Zika virus affects cortical progenitor cells [16], as well as postmigrational neurons of the neocortex, leading to apoptosis [6, 12], what would result in arrested differentiation of the neocortex [9].

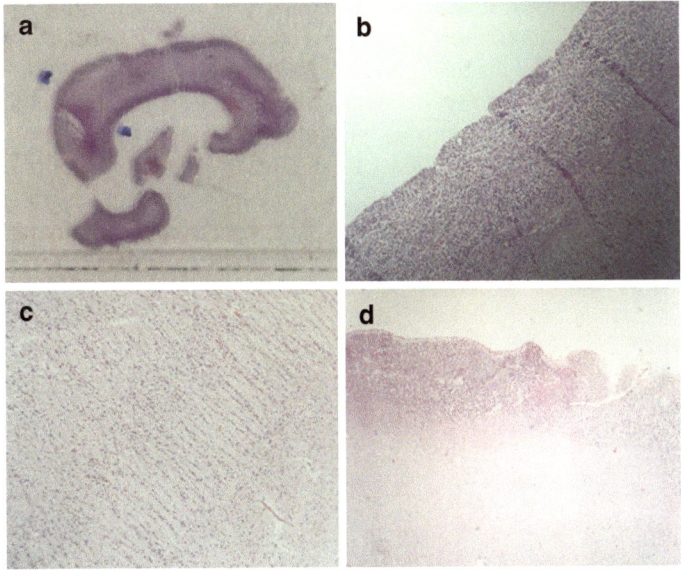

FIGURE 5 Hematoxylin-eosin-stained brain slide (**a**) showing neuroblasts concentrated at the periventricular and cortical areas. In a higher magnification, high concentration of neuroblasts can be identified at the germinative matrix (periventricular area) (**b**); decreased number of migrating neuroblasts at the radial glia (**c**); and irregularly distributed neuroblasts at the abnormally organized cortex (**d**), which does not have proper gyration and sulcation

In other organs, there is no direct Zika-related damage, and there is inconsistent viral detection by immuno-histochemistry and real-time protein chain reaction (RT-PCR), findings that support the specificity of Zika virus for the central nervous system [6, 9, 11, 14]. The only other tissue in which direct viral damage has been detected is the placenta [7–10].

The viral damage is identified through its sequelae in the brain tissue; signs of current viral or inflammatory activity are not seen in the brain slides. There are activated microglia and macrophages, but inflammation is not prominent [8, 12]. In addition to its specific cytopathic effect to the brain, this lack of inflammatory response distinguishes the Zika virus infection from other central nervous system congenital infections that cause similar manifestations [8, 14].

Lesions are diffuse at the neocortex, calcifications being the most characteristic finding [7–10]. Calcifications are located in the cortical and subcortical white matter, and they can have coarse aspect, signalizing a big area of destruction (Fig. 6a), or filamentous, granular, and neuron shapes [9], contributing to the idea of calcium deposition in areas of neuronal lesions (Fig. 6b). Early mineralization was described

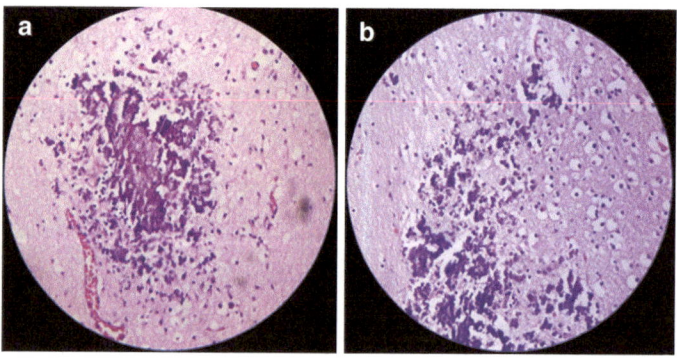

FIGURE 6 Hematoxylin-eosin-stained brain slides show areas of calcifications, which can have a coarse aspect of calcium deposition (**a**) and can also present neuron-shaped aspect (**b**)

in association with the apoptotic neurons [6], further supporting this hypothesis.

Microscopically, pachygyria is characterized by altered thickness of cortical layers and presence of a cell-sparse zone between inner and outer layers, while polymicrogyria consists of deranged cortical layers [17]. This differentiation is interesting in lights of the report that polymicrogyria seen on MRI can in fact be absent in histopathological analysis [12]. It is proposed that focal glioneuronal outbursts into the subarachnoid space, in association with calcifications and heterotopia, could result in the radiological appearance of polymicrogyria [12]. Therefore, the definitive distinction between pachygyria and polymicrogyria is done through histopathology, although MRI has been shown to differentiate well these types of malformation in most cases.

The white matter is severely reduced in volume, with extensive axonal rarefaction [6]. Gliosis is not a characteristic finding; however, subependymal gliosis, shown in Fig. 7, is seen in some cases.

Organization of the cerebellum and brainstem can be preserved, although hypoplasia, Wallerian degeneration and calcifications can also occur [12, 13]. When the cerebellum is

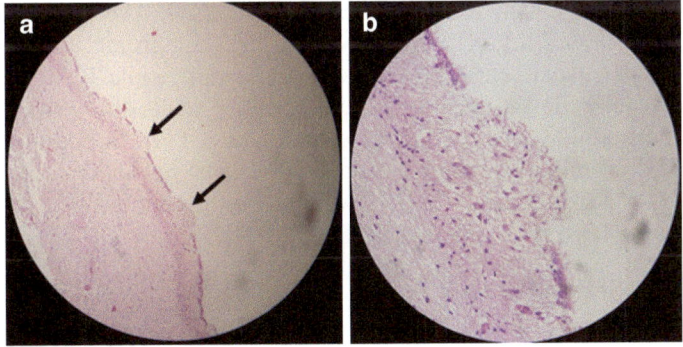

Figure 7 Hematoxylin-eosin-stained brain slide (a) shows the ependymal cell layer interrupted in two points by subependymal gliosis (*black arrows*), better seen in a higher magnification (b)

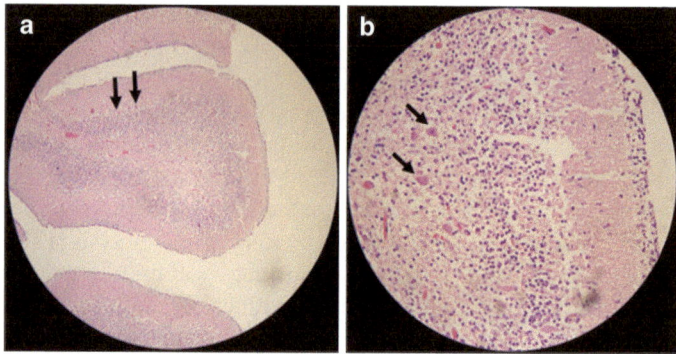

Figure 8 Hematoxylin-eosin-stained cerebellum slide (**a**) shows a folium decreased in size, with few cells in the Purkinje layer (*black arrows*) and decreased granular layer. In a higher magnification (**b**), Purkinje cells can be identified outside the area of their layer (*black arrows*)

hypoplastic, the folia are decreased in size and sulci are larger, as seen in Fig. 8. In our experience, the Purkinje cell layer is not identified through the entire cortex of the folia, and the granular layer is diminished (Fig. 8a). Some Purkinje cells can be found outside the area in which they should be located (Fig. 8b).

Despite of the diffuse and severe pattern of the lesions found especially in the neocortex, some areas of the brain of fetuses and infants with congenital Zika syndrome are less damaged or even well preserved. For example, while Wallerian degeneration is present in the descending tracts of the spinal cord, the ascending tracts are not damaged [9]. In addition, hippocampus, entorhinal cortex, subcortical nuclei, thalamus and limbic regions are sites with well-differentiated neurons and preserved organization [6, 12].

References

1. de Fatima Vasco Aragao M, Linden V v d, Brainer-Lima AM, et al. Clinical features and neuroimaging (CT and MRI) findings in presumed Zika virus related congenital infection and microcephaly: retrospective case series study. BMJ. 2016;353:i1901.

2. Carvalho FHC, Cordeiro KM, Peixoto AB, Tonni G, Moron AF, Feitosa FEL, Feitosa HN, Araujo Júnior E. Associated ultrasonographic findings in fetuses with microcephaly because of suspected Zika virus (ZIKV) infection during pregnancy. Prenat Diagn. 2016;36:882–7.

3. Hazin AN, Poretti A, Cruz DDCS, et al. Computed tomographic findings in microcephaly associated with Zika virus. N Engl J Med. 2016;374(22):2193–5.

4. Oliveira Melo AS, Malinger G, Ximenes R, Szejnfeld PO, Alves Sampaio S, Bispo de Filippis AM. Zika virus intrauterine infection causes fetal brain abnormality and microcephaly: tip of the iceberg? Ultrasound Obstet Gynecol. 2016;47:6–7.

5. Soares de Oliveira-Szejnfeld P, Levine D, Melo AS d O, et al. Congenital brain abnormalities and Zika virus: what the radiologist can expect to see prenatally and postnatally. Radiology. 2016;281:203–18.

6. Driggers RW, Ho C-Y, Korhonen EM, et al. Zika virus infection with prolonged maternal viremia and fetal brain abnormalities. N Engl J Med. 2016;374:2142–51.

7. Martines RB, Bhatnagar J, Keating MK, et al. Notes from the field: evidence of Zika virus infection in brain and placental tissues from two congenitally infected newborns and two fetal losses—Brazil, 2015. MMWR Morb Mortal Wkly Rep. 2016;65:159–60.

8. Martines RB, Bhatnagar J, de Oliveira Ramos AM, et al. Pathology of congenital Zika syndrome in Brazil: a case series. Lancet Lond Engl. 2016;388:898–904.

9. Mlakar J, Korva M, Tul N, et al. Zika virus associated with microcephaly. N Engl J Med. 2016;374:951–8.

10. Noronha L d, Zanluca C, Azevedo MLV, Luz KG, Santos CNDD. Zika virus damages the human placental barrier and presents marked fetal neurotropism. Mem Inst Oswaldo Cruz. 2016;111:287–93.

11. Sarno M, Sacramento GA, Khouri R, et al. Zika virus infection and stillbirths: a case of hydrops fetalis, hydranencephaly and fetal demise. PLoS Negl Trop Dis. 2016;10:e0004517.

12. Štrafela P, Vizjak A, Mraz J, Mlakar J, Pižem J, Tul N, Županc TA, Popović M. Zika virus-associated micrencephaly: a thorough description of neuropathologic findings in the fetal central nervous system. Arch Pathol Lab Med. 2016;141(1):73–81. doi:10.5858/arpa.2016-0341-SA.

13. Melo ASO, Aguiar RS, Amorim MMR, et al. Congenital Zika Virus Infection: Beyond Neonatal Microcephaly. JAMA Neurol. 2016;73:1407–16.
14. Schwartz DA. Autopsy and postmortem studies are concordant: pathology of Zika virus infection is neurotropic in fetuses and infants with microcephaly following transplacental transmission. Arch Pathol Lab Med. 2016;141(1):68–72. doi:10.5858/arpa.2016-0343-OA.
15. Linden V v d, Filho ELR, Lins OG, et al. Congenital Zika syndrome with arthrogryposis: retrospective case series study. BMJ. 2016;354:i3899.
16. Tang H, Hammack C, Ogden SC, et al. Zika virus infects human cortical neural progenitors and attenuates their growth. Cell Stem Cell. 2016;18:587–90.
17. Barkovich AJ, Gressens P, Evrard P. Formation, maturation, and disorders of brain neocortex. Am J Neuroradiol. 1992;13:423–46.

Index